LECTURES ON GEOMETRY

Series Editor

N. M. J. WOODHOUSE

Lectures on Geometry

Edited by

N. M. J. WOODHOUSE

President, Clay Mathematics Institute

OXFORD
UNIVERSITY PRESS

OXFORD
UNIVERSITY PRESS

Great Clarendon Street, Oxford, OX2 6DP,
United Kingdom

Oxford University Press is a department of the University of Oxford.
It furthers the University's objective of excellence in research, scholarship,
and education by publishing worldwide. Oxford is a registered trade mark of
Oxford University Press in the UK and in certain other countries

Published in the United States of America by Oxford University Press
198 Madison Avenue, New York, NY 10016, United States of America

British Library Cataloguing in Publication Data
Data available

Library of Congress Control Number: 2016943647

ISBN 978–0–19–878491–3

Printed and bound by
CPI Group (UK) Ltd, Croydon, CR0 4YY

Preface

This volume contains a collection of papers based on lectures delivered by distinguished mathematicians at Clay Mathematics Institute events over the past few years. It is intended to be the first in an occasional series of volumes of CMI lectures. Although not explicitly linked, the topics in this inaugural volume have a common flavour and a common appeal to all who are interested in recent developments in geometry. They are intended to be accessible to all who work in this general area, regardless of their own particular research interests.

Two Lectures on the Jones Polynomial and Khovanov Homology

Edward Witten

Edward Witten works at the Institute for Advanced Study at Princeton. He is one of the leading figures in contemporary theoretical physics. His chapter is based on two lectures he gave at the Clay Research Conference in 2013. It surveys the groundbreaking work of Witten and his collaborators in fitting Khovanov homology into a quantum field theory framework. In the abstract of his contribution, Witten says: 'I describe a gauge theory approach to understanding quantum knot invariants as Laurent polynomials in a complex variable q. The two main steps are to reinterpret three-dimensional Chern–Simons gauge theory in four-dimensional terms and then to apply electric–magnetic duality. The variable q is associated to instanton number in the dual description in four dimensions.' This hardly does justice to the extraordinary range of ideas and techniques from mathematics and theoretical physics on which his lectures drew in his journey from an elementary starting point in the classical theory of knots. The second lecture was delivered in the Number Theory and Physics workshop at the conference. It takes the story further, describing how Khovanov homology can emerge upon adding a fifth dimension. Along the way, Witten describes many significant new ideas, such as the Kapustin–Witten equations (important in geometric Langlands) and a new approach to evaluating some Feynman integrals via complexification. Witten's approach is very natural, and especially attractive to a geometer, using Picard–Lefschetz theory in an essential way.

Elementary Knot Theory

Marc Lackenby

Marc Lackenby is a Professor of Mathematics at Oxford, with special interests in geometry and topology in three dimensions. His chapter is partly based on a lecture at the Clay Research Conference in 2012. It focuses on identifying some fundamental problems in knot theory that are easy to state but that remain unsolved. A survey of this very active field is given to place these problems into context. Because the tools that are used in knot theory are so diverse, the chapter highlights connections with many other fields of mathematics, including hyperbolic geometry, the theory of computational complexity, geometric group theory (a large area that connects with Bridson's chapter) and Khovanov homology (the subject of Witten's chapter).

Cube Complexes, Subgroups of Mapping Class Groups and Nilpotent Genus

Martin R. Bridson

Martin Bridson is the Whitehead Professor of Mathematics at Oxford, well known for his work in geometric group theory. His contribution is based on the lecture he gave as a Clay Senior Scholar at the Park City Mathematics Institute in 2012. This event is organized each summer by the Institute for Advanced Study at Princeton and is supported by the Clay Mathematics Institute through the appointment of Clay Senior Scholars. The PCMI Scholars provide mathematical leadership for the summer programmes and deliver lectures addressed to a wide mathematical audience. Bridson's chapter focuses on two recent sets of results of his, one on mapping class groups of surfaces and the other on nilpotent genera of groups, both of which illuminate extreme behaviour among finitely presented groups. It provides an extremely useful and readable introduction to an important and lively area.

Polyfolds and Fredholm Theory

Helmut Hofer

Helmut Hofer is a member of the Institute for Advanced Study at Princeton. He has played a major part in the development of symplectic topology. The original version of this important and previously unpublished chapter was written following the Clay Research Conference in 2008, at which Hofer spoke. Since then it has been extended and revised to bring it up to date. The chapter discusses generalized Fredholm theory in polyfolds, an area in which Hofer is a leading figure, with a focus on a particular topic—stable maps—that has a close connection to Gromov–Witten theory. This selection allows Hofer to set his chapter within a broad context. His excellent and full introduction makes accessible the very detailed exposition that follows.

Maps, Sheaves and $K3$ Surfaces

Rahul Pandharipande

Rahul Pandharipande works at ETH Zürich. He is well known for his work with Okounkov, Nekrasov and Maulik on Gromov–Witten theory and Donaldson–Thomas invariants, for which he received a Clay Research Award from CMI in 2013. Pandharipande's chapter also arises from a lecture delivered at the Clay Research Conference in 2008, in which he reviewed his work and that of his collaborators on recent progress in understanding curve counting (Gromov–Witten theory and its cousins) in higher dimensions. Gromov–Witten theory is notoriously hard and is only fully understood in dimensions 0 and 1. Pandharipande describes progress in dimensions 2 and 3. The chapter concisely describes a wide variety of important geometric ideas and useful techniques. It ends by bringing the story up to date with a brief account of the successful proofs of some of the principal conjectures covered in the original lecture.

N. M. J. WOODHOUSE
Clay Mathematics Institute

Contents

List of Contributors

MARTIN R. BRIDSON
Mathematical Institute
University of Oxford
Andrew Wiles Building
Radcliffe Observatory Quarter
Woodstock Road
Oxford OX2 6GG, UK

HELMUT H. W. HOFER
School of Mathematics
Institute for Advanced Study
Einstein Drive
Princeton, NJ 08540, USA

MARC LACKENBY
Mathematical Institute
University of Oxford
Andrew Wiles Building
Radcliffe Observatory Quarter
Woodstock Road
Oxford OX2 6GG, UK

RAHUL PANDHARIPANDE
Department of Mathematics
ETH Zürich
Rämistrasse 101
8092 Zürich
Switzerland

EDWARD WITTEN
School of Natural Sciences
Institute for Advanced Study
Einstein Drive
Princeton, NJ 08540, USA

N. M. J. WOODHOUSE
CMI President's Office
Andrew Wiles Building
Radcliffe Observatory Quarter
Woodstock Road
Oxford OX2 6GG, UK

1
Two Lectures on the Jones Polynomial and Khovanov Homology

EDWARD WITTEN

1.1 Lecture One

The Jones polynomial is a celebrated invariant of a knot (or link) in ordinary three-dimensional space, originally discovered by V. F. R. Jones roughly thirty years ago as an offshoot of his work on von Neumann algebras [1]. Many descriptions and generalizations of the Jones polynomial were discovered in the years immediately after Jones's work. They more or less all involved statistical mechanics or two-dimensional mathematical physics in one way or another—for example, Jones's original work involved Temperley–Lieb algebras of statistical mechanics. I do not want to assume that the Jones polynomial is familiar to everyone, so I will explain one of the original definitions.

For brevity, I will describe the "vertex model" (see [2] and also [3], p. 125). One projects a knot to $\mathbb{R}^2$ in such a way that the only singularities are simple crossings and so that the height function has only simple local maxima and minima (Fig. 1.1). One labels the intervals between crossings, maxima and minima by a symbol + or –. One sums over all possible labelings of the knot projection with simple weight functions given in Figs. 1.2 and 1.3. The weights are functions of a variable q. After summing over all possible labelings and weighting each labeling by the product of the weights attached to its crossings, maxima and minima, one arrives at a function of q. The sum turns out to be an invariant of a framed knot.[1] This invariant is a Laurent polynomial in q (times a fixed fractional power of q that depends on the framing). It is known as the Jones polynomial.

Clearly, given the rules stated in the figures, the Jones polynomial for a given knot is completely computable by a finite (but exponentially long) algorithm. The rules, however, seem to have come out of thin air. Topological invariance is not obvious and is proved by checking Reidemeister moves.

Lectures on Geometry. Edward Witten, Marc Lackenby, Martin R. Bridson, Helmut Hofer and Rahul Pandharipande.
© Oxford University Press 2017. Published 2017 by Oxford University Press.

Figure 1.1 A knot in $\mathbb{R}^3$—in this case a trefoil knot—projected to the plane $\mathbb{R}^2$ in a way that gives an immersion with only simple crossings and such that the height function (the vertical coordinate in the figure) has only simple local maxima and minima. In this example, there are three crossings (each of which contributes two crossing points, one on each branch) and two local minima and maxima, making a total of $3 \cdot 2 + 2 + 2 = 10$ exceptional points. Omitting those points divides the knot into 10 pieces that can be labeled by symbols + or −, so the vertex model for this projection expresses the Jones polynomial of the trefoil knot as a sum of 2^{10} terms.

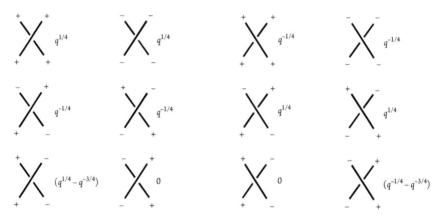

Figure 1.2 The weights of the vertex model for a simple crossing of two strands. (The weights for configurations not shown are 0.)

Other descriptions of the Jones polynomial were found during the same period, often involving mathematical physics. The methods involved statistical mechanics, braid group representations, quantum groups, two-dimensional conformal field theory and more. One notable fact was that conformal field theory can be used [4] to generalize the constructions of Jones to the choice of an arbitrary simple Lie group[2] $G^\vee$ with a labeling of a knot (or of each component of a link) by an irreducible representation $R^\vee$ of $G^\vee$.

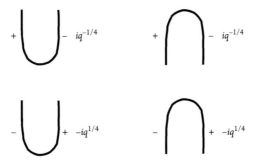

Figure 1.3 The weights of the vertex model for a local maximum or minimum of the height function. (Weights not shown are 0.)

The original Jones polynomial is the case that $G^\vee = \mathrm{SU}(2)$ and $R^\vee$ is the two-dimensional representation.

With these and other clues, it turned out [5] that the Jones polynomial can be described in three-dimensional quantum gauge theory. Here we start with a compact simple gauge group $G^\vee$ (to avoid minor details, we take $G^\vee$ to be connected and simply connected) and a trivial[3] $G^\vee$-bundle $E^\vee \to W$, where W is an oriented three-manifold. Let A be a connection on $E^\vee$. The only gauge-invariant function of A that we can write by integration over W of some local expression, assuming no structure on W except an orientation, is the Chern–Simons function

$$\mathrm{CS}(A) = \frac{1}{4\pi} \int_W \mathrm{Tr}\left(A \wedge dA + \frac{2}{3} A \wedge A \wedge A \right). \tag{1.1}$$

Even this function is only gauge-invariant modulo a certain fundamental period. In (1.1), Tr is an invariant and non-degenerate quadratic form on the Lie algebra of $G^\vee$, normalized so that $\mathrm{CS}(A)$ is gauge-invariant mod $2\pi\,\mathbb{Z}$. For $G^\vee = \mathrm{SU}(n)$ (for some $n \geq 2$), we can take Tr to be the trace in the n-dimensional representation.

The Feynman path integral is now formally an integral over the infinite-dimensional space U of connections:

$$Z_k(W) = \frac{1}{\mathrm{vol}} \int_U DA \, \exp[ik\mathrm{CS}(A)]. \tag{1.2}$$

This is a basic construction in quantum field theory, though unfortunately challenging to understand from a mathematical point of view. Here k has to be an integer since $\mathrm{CS}(A)$ is only gauge-invariant modulo $2\pi\,\mathbb{Z}$. $Z_k(W)$ is defined with no structure on W except an orientation, so it is an invariant of the oriented three-manifold W. (Here and later, I ignore some details. W actually has to be "framed," as one learns if one follows the logic of "renormalization theory." Also, formally, vol is the volume of the infinite-dimensional group of gauge transformations.)

To include a knot—that is, an embedded oriented circle $K \subset W$—we make use of the *holonomy* of the connection A around W, which we denote by $\mathrm{Hol}(A, K)$. We pick an irreducible representation $R^\vee$ of $G^\vee$ and define

$$\mathcal{W}_{R^\vee}(K) = \mathrm{Tr}_{R^\vee} \mathrm{Hol}_K(A) = \mathrm{Tr}_{R^\vee} P \exp\left(-\oint_K A\right), \tag{1.3}$$

where the last expression is the way that physicists often denote the trace of the holonomy. In the context of quantum field theory, the trace of the holonomy is usually called the Wilson loop operator. Then we define a natural invariant of the pair W, K:

$$Z_k(W; K, R^\vee) = \frac{1}{\mathrm{vol}} \int_U DA \exp[ikCS(A)] \mathcal{W}_{R^\vee}(K). \tag{1.4}$$

(Again, framings are needed.)

If we take $G^\vee$ to be $SU(2)$ and $R^\vee$ to be the two-dimensional representation, then $Z_k(W; K, R^\vee)$ turns out to be the Jones polynomial, evaluated at[4]

$$q = \exp\left(\frac{2\pi i}{k+2}\right). \tag{1.5}$$

This statement is justified by making contact with two-dimensional conformal field theory, via the results of [4]. For a particularly direct way to establish the relation to the Knizhnik–Zamolodchikov equations of conformal field theory, see [6]. This relationship between three-dimensional gauge theory and two-dimensional conformal field theory has also been important in condensed matter physics, in studies of the quantum Hall effect and related phenomena.

This approach has more or less the opposite virtues and drawbacks to those of the standard approaches to the Jones polynomial. No projection to a plane is chosen, so topological invariance is obvious (modulo standard quantum field theory machinery), but it is not clear how much one will be able to compute. In other approaches, like the vertex model, there is an explicit finite algorithm for computation, but topological invariance is obscure.

Despite the manifest topological invariance of this approach to the Jones polynomial, there were at least two things that many knot theorists did not like about it. One was simply that the framework of integration over function spaces—though quite familiar to physicists—is difficult to understand mathematically. (A version of this problem is one of the Clay Millennium Problems.) The second is that this method did not give a clear approach to understanding why the usual quantum knot invariants are Laurent polynomials in q. This method, in its original form, gave a definition of the knot invariants only for integer k, and did not explain the existence of an analytic continuation to a function of a complex variable q, let alone the fact that the analytically continued functions are Laurent polynomials. From some points of view, the fact that the invariants are Laurent polynomials is considered sufficiently important that it is part of the name "Jones polynomial." Other approaches to the Jones polynomial—such as the vertex model that we started with—do not obviously give a topological invariant but do obviously give a Laurent polynomial.

Actually, for most three-manifolds, the answer that comes from the gauge theory is the right one. It is special to knots in $\mathbb{R}^3$ that the natural variable is $q = \exp[2\pi i/(k+2)]$ rather than k. The quantum knot invariants on a general three-manifold W are naturally defined only for an integer k and do not have natural analytic continuations to functions of[5] q. This has been the traditional understanding: the gauge theory gives directly a good understanding on a general three-manifold W, but if one wants to understand from three-dimensional gauge theory some of the special things that happen for knots in $\mathbb{R}^3$, one has to begin by relating the gauge theory to one of the other approaches, for instance via conformal field theory.

However, a little over a decade ago, two developments gave clues that there should be another explanation. One of these developments was Khovanov homology, which will be the topic of the second lecture. The other development, which started at roughly the same time, was the "volume conjecture" [7–12]. What I will explain in this lecture started with an attempt to understand the volume conjecture. I should stress that I have not succeeded in finding a quantum field theory explanation for the volume conjecture.[6] However, just understanding a few preliminaries concerning the volume conjecture led to a new point of view on the Jones polynomial. This is what I aim to explain. Since this is the case, I will actually not give a precise statement of the volume conjecture.

To orient ourselves, let us just ask how the basic integral

$$Z_k(W) = \frac{1}{\text{vol}} \int_U DA \, \exp[ik\text{CS}(A)] \tag{1.6}$$

behaves for large k. It is an infinite-dimensional analog of a finite-dimensional oscillatory integral such as the one that defines the Airy function

$$F(k;t) = \int_{-\infty}^{\infty} dx \, \exp[ik(x^3 + tx)], \tag{1.7}$$

where we assume that k and t are real. Taking $k \to \infty$ with fixed t, the integral vanishes exponentially fast if there are no real critical points $(t > 0)$ and is a sum of oscillatory contributions of real critical points if there are any $(t < 0)$. The same logic applies to the infinite-dimensional integral for $Z_k(W)$. The critical points of $\text{CS}(A)$ are flat connections, corresponding to homomorphisms $\rho : \pi_1(W) \to G$, so the asymptotic behavior of $Z_k(W)$ for large k is given by a sum of oscillatory contributions associated to such homomorphisms. (This has been shown explicitly in examples [14, 15].)

The volume conjecture arises if we specialize to knots in $\mathbb{R}^3$, so that—as one knows from any approach to the Jones polynomial other than that via Chern–Simons gauge theory—k does not have to be an integer. Usually the case $G^\vee = \text{SU}(2)$ is assumed and we let $R^\vee$ be the n-dimensional representation of $\text{SU}(2)$. The corresponding knot invariant is called the colored Jones polynomial. We take $k \to \infty$ through non-integer values, with fixed k/n. A choice that is sufficient to illustrate the main points is to set $k = k_0 + n$, where k_0 is a fixed complex number and we take $n \to \infty$ (through integer values). The behavior of the colored Jones polynomial in this limit has been studied for a variety of knots, using approaches to the knot invariants in which there is no restriction to integer k, for example the approach via quantum groups. Very interesting results have emerged from this work [7–12]. Trying to understand these results via path integrals was the motivation for what I am describing in this lecture.

What emerged from study of the limit $n \to \infty$ with $k = k_0 + n$ is very suggestive of Chern–Simons gauge theory, but with a crucial twist. In examples that have been studied, the large-n behavior can be interpreted in terms of a sum of critical points of the Chern–Simons path integral, but now these are *complex* critical points. By a complex critical point, I mean simply a critical point of the analytic continuation of the function $CS(A)$.

We make this analytic continuation simply by replacing the Lie group $G^\vee$ with its complexification $G_{\mathbb{C}}^\vee$, replacing the $G^\vee$-bundle $E^\vee \to W$ with its complexification, which is a $G_{\mathbb{C}}^\vee$ bundle $E_{\mathbb{C}}^\vee \to W$, and replacing the connection A on $E^\vee$ by a connection $\mathcal{A}$ on $E_{\mathbb{C}}^\vee$, which we can think of as a complex-valued connection. Once we do this, the function $CS(A)$ on the space U of connections on $E^\vee$ can be analytically continued to a holomorphic function $CS(\mathcal{A})$ on $\mathcal{U}$, the space of connections on $E_{\mathbb{C}}^\vee$. This function is defined by the "same formula" with A replaced by $\mathcal{A}$:

$$CS(\mathcal{A}) = \frac{1}{4\pi} \int_W \mathrm{Tr}\left(\mathcal{A} \wedge \mathrm{d}\mathcal{A} + \frac{2}{3} \mathcal{A} \wedge \mathcal{A} \wedge \mathcal{A} \right). \tag{1.8}$$

On a general three-manifold W, a critical point of $CS(\mathcal{A})$ is simply a complex-valued flat connection, corresponding to a homomorphism $\rho : \pi_1(W) \to G_{\mathbb{C}}^\vee$.

In the case of the volume conjecture with $W = \mathbb{R}^3$, the fundamental group is trivial, but we are supposed to also include a holonomy or Wilson loop operator $\mathcal{W}_{R^\vee}(K) = \mathrm{Tr}_{R^\vee} \mathrm{Hol}_K(A)$, where $R^\vee$ is the n-dimensional representation of $SU(2)$. When we take $k \to \infty$ with fixed k/n, this holonomy factor affects what we should mean by a critical point.[7] A full explanation would take us too far afield, and instead I will just give the answer: the right notion of a complex critical point for the colored Jones polynomial is a homomorphism $\rho : \pi_1(W \backslash K) \to G_{\mathbb{C}}^\vee$, with a monodromy around K whose conjugacy class is determined by the ratio n/k. What is found in work on the "volume conjecture" is that (in examples that have been studied) the colored Jones polynomial for $k \to \infty$ with fixed n/k is determined by such a complex critical point.

Physicists know about various situations (involving "tunneling" problems) in which a path integral is dominated by a complex critical point, but usually this is a complex critical point that makes an exponentially small contribution. There is a simple reason for this. Usually in quantum mechanics, one is computing a probability amplitude. Since probabilities cannot be bigger than 1, the contribution of a complex critical point to a probability amplitude can be exponentially small but it cannot be exponentially large. What really surprised me about the volume conjecture is that, for many knots (knots with hyperbolic complement in particular), the dominant critical point makes an exponentially *large* contribution. In other words, the colored Jones polynomial is a sum of oscillatory terms for $n \to \infty, k = k_0 + n$ if k_0 is an integer, but it grows exponentially in this limit as soon as k_0

is not an integer. (Concretely, this is because $kCS(\mathcal{A})$ evaluated at the appropriate critical point has a negative imaginary part, so $\exp[ikCS(\mathcal{A})]$ grows exponentially for large k.)

There is no contradiction with the statement that quantum mechanical probability amplitudes cannot be exponentially large, because as soon as k_0 is not an integer, we are no longer studying a physically sensible quantum mechanical system. But it seemed puzzling that making k_0 non-integral, even if still real, can change the large-n behavior so markedly. However, it turns out that a simple one-dimensional integral can do the same thing:

$$I(k, n) = \int_0^{2\pi} \frac{d\theta}{2\pi} e^{ik\theta} e^{2in\sin\theta}.$$ (1.9)

We want to think of k and n as analogs of the integer-valued parameters in Chern–Simons gauge theory that we call by the same names. (In our model problem, k is naturally an integer, but there is no good reason for n to be an integer. So the analogy is not perfect.) If one takes k, n to infinity with a fixed (real) ratio and maintaining the integrality of k, then the integral $I(k, n)$ has an oscillatory behavior, dominated by the critical points of the exponent $f = k\theta + 2n\sin\theta$, if k/n is such that there are critical points for real θ. Otherwise, the integral vanishes exponentially fast for large k.

Now, to imitate the situation considered in the volume conjecture, we want to analytically continue away from integer values of k. The integral $I(k, n)$ obeys Bessel's equation (as a function of n) for any integer k. We want to think of Bessel's equation as the analog of the "Ward identities" of quantum field theory, so in the analytic continuation of $I(k, n)$ away from integer k, we want to preserve Bessel's equation. The proof of Bessel's equation involves integration by parts, so it is important that we are integrating all the way around the circle and that the integrand is continuous and single-valued on the circle. That is why k has to be an integer.

The analytic continuation of $I(k, n)$, preserving Bessel's equation, was known in the nineteenth century. We first set $z = e^{i\theta}$, so our integral becomes

$$I(k, n) = \oint \frac{dz}{2\pi i} z^{k-1} \exp[n(z - z^{-1})].$$ (1.10)

Here the integral is over the unit circle in the z-plane. At this point, k is still an integer. We want to get away from integer values while still satisfying Bessel's equation. If $\mathrm{Re}\, n > 0$, this can be done by switching to the integration cycle shown in Fig. 1.4.

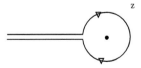

Figure 1.4 The contour used in analytic continuation of the solution of Bessel's equation.

The integral on the new cycle converges (if $\operatorname{Re} n > 0$), and it agrees with the original integral on the circle if k is an integer, since the extra parts of the cycle cancel. But the new cycle gives a continuation away from integer k, still obeying Bessel's equation. There is no difficulty in the integration by parts used to prove Bessel's equation, since the integral on the chosen cycle is rapidly convergent at infinity.

How does the integral on the new cycle behave in the limit $k, n \to \infty$ with fixed k/n? If k is an integer and n is real, then the integral is oscillatory or exponentially damped, as I have stated before, depending on the ratio k/n. But as soon as k is not an integer (even if k and n remain real), the large-k behavior with fixed k/n is one of exponential growth, for a certain range of k/n, rather as is found for the colored Jones polynomial. Unfortunately, even though it is elementary, a full explanation of this statement would involve a bit of a digression. (Details can be found, for example, in [13], Section 3.5.) Here I will just explain the technique that one can use to make this analysis, since this will show the technique that we will follow in taking a new look at the Jones polynomial.

We are trying to do an integral of the generic form

$$\int_\Gamma \frac{dz}{2\pi i z} \exp[kF(z)], \tag{1.11}$$

where $F(z)$ is a holomorphic function and Γ is a cycle, possibly not compact, on which the integral converges. In our case,

$$F(z) = \log z + \lambda(z - z^{-1}), \qquad \lambda = n/k. \tag{1.12}$$

We note that because of the logarithm, $F(z)$ is multivalued. To do the analysis properly, we should work on a cover of the punctured z-plane parametrized by $w = \log z$ on which F is single-valued:

$$F(w) = w + \lambda(e^w - e^{-w}). \tag{1.13}$$

The next step is to find a useful description of all possible cycles on which the desired integral, which now is

$$\int_\Gamma \frac{dw}{2\pi i} \exp[kF(w)], \tag{1.14}$$

converges.

Morse theory gives an answer to this question. We consider the function $h(w, \bar{w}) = \operatorname{Re}[kF(w)]$ as a Morse function. Its critical points are simply the critical points of the holomorphic function F, and so in our example they obey

$$1 + \lambda(e^w + e^{-w}) = 0. \tag{1.15}$$

The key step is now the following. To every critical point p of F, we can define an integration cycle Γ_p, called a Lefschetz thimble, on which the integral we are trying to do converges. Moreover, the Γ_p give a basis of integration cycles on which this integral converges, since they give a basis of the homology of the w-plane relative to the region with $h \to -\infty$. (We assume that the critical points of F are all non-degenerate, as is the case in our example. Also, we assume that F is sufficiently generic that the equation (1.16) introduced momentarily has no solutions interpolating from one critical point at $t = -\infty$ to another at $t = +\infty$. If F varies as a function of some parameters, then in real codimension 1, such interpolating solutions do appear; there is then a Stokes phenomenon—a jumping in the basis of the relative homology given by the Γ_p.)

In fact, since h is the real part of a holomorphic function, its critical points are all saddle points, not local maxima or minima. The Lefschetz thimble associated to a given critical point p is defined by "flowing down" from p (Fig. 1.5), via the gradient flow equation of Morse theory. We could use any complete Kahler metric on the w-plane in defining this equation, but we may as well use the obvious flat metric $ds^2 = |dw|^2$. The gradient flow equation is then

$$\frac{dw}{dt} = -\frac{\partial h}{\partial \overline{w}}, \qquad (1.16)$$

where t is a new "time" coordinate. The Lefschetz thimble Γ_p associated to a critical point p is defined as the space of all values at $t = 0$ of solutions of the flow equation on the semi-infinite interval $(-\infty, 0]$ that start at p at $t = -\infty$. For example, p itself is contained in Γ_p, because it is the value at $t = 0$ of the trivial solution of the flow equation that is equal to p for all times. A non-constant solution that approaches p for $t \to -\infty$ is exponentially close to p for large negative t. The coefficient of the exponentially small term in a particular solution determines how far the flow reaches by time $t = 0$ and therefore what point on Γ_p is represented by this particular flow.

Γ_p is not compact, but the integral

$$I_p = \int_{\Gamma_p} \frac{dw}{2\pi i} \exp[kF(w)] \qquad (1.17)$$

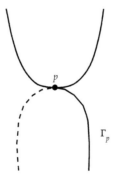

Figure 1.5 The Lefschetz thimble associated to a critical point. The critical point is a saddle point and the thimble is the union of downward flows that start at this saddle.

converges, since $h = \text{Re}[kF(w)]$ goes to $-\infty$ at infinity along Γ_p. Moreover, when restricted to Γ_p, h has a unique maximum, which is at the point p. This statement leads to a straightforward answer for the large-k behavior of the integral I_p:

$$I_p \sim \exp[kF(p)]\left(c_0 k^{-1/2} + c_1 k^{-3/2} + \dots\right), \tag{1.18}$$

where the coefficients $c_0, c_1, \dots$ in the asymptotic expansion can be computed by classical methods.

Any other cycle Γ on which the integral converges can be expanded as a linear combination of the Lefschetz thimbles:

$$\Gamma = \sum_p a_p \Gamma_p, \qquad a_p \in \mathbb{Z}. \tag{1.19}$$

After computing the integers a_p, it is straightforward to determine the large-k asymptotics of an integral

$$\int_\Gamma \frac{dw}{2\pi i} \exp[kF(w)]. \tag{1.20}$$

It is simply given by the contributions of those critical points p for which $h(p)$ is maximal under the condition that $a_p \neq 0$. Applying this procedure to our example related to the Bessel function, we get the answer that I claimed before: this integral has an asymptotic behavior similar to that of the colored Jones polynomial. The limit $n \to \infty, k = k_0 + n$ is quite different depending on whether k_0 is an integer. (Concretely, if k_0 is not an integer, the large-n behavior is dominated by two Lefschetz thimbles whose contributions cancel if k_0 is an integer.)

At this stage, I hope it is fairly clear what we should do to understand the analytic continuation to non-integer k of the quantum invariants of knots in $\mathbb{R}^3$, and also to understand the asymptotic behavior of the colored Jones polynomial that is related to the volume conjecture. We should define Lefschetz thimbles in the space $\mathcal{U}$ of complex-valued connections, or, more precisely, in a cover of this space on which $CS(\mathcal{A})$ is single-valued, and in the gauge theory definition of the Jones polynomial, we should replace the integral over the space U of real connections with a sum of integrals over Lefschetz thimbles.

However, it probably is not clear that this will actually lead to a useful new viewpoint on the Jones polynomial. This depends on a few additional facts. To define the Lefschetz thimbles that we want, we need to consider a gradient flow equation on the infinite-dimensional space $\mathcal{U}$ of complex-valued connections, with $\text{Re}[ikCS(\mathcal{A})]$ as a Morse function.[8] Actually, I want to first practice with the case of gradient flow on the infinite-dimensional space U of real connections (on a $G^\vee$-bundle $E^\vee \to W$, W being a three-manifold) with the Morse function being the real Chern–Simons function $CS(A)$. This case is important in Floer theory of three-manifolds and in Donaldson theory of smooth four-manifolds, so it is relatively familiar. A Riemannian metric on W induces a Riemannian metric on U by

$$|\delta A|^2 = -\int_W \mathrm{Tr}\,\delta A \wedge \star \delta A, \tag{1.21}$$

where $\star = \star_3$ is the Hodge star operator acting on differential forms on W. We will use this metric in defining a gradient flow equation on U, with Morse function $\mathrm{CS}(A)$.

The flow equation will be a differential equation on a four-manifold $M = W \times \mathbb{R}$, where $\mathbb{R}$ is parametrized by the "time"; one can think of the flow as evolving a three-dimensional connection in "time." Concretely, the flow equation is

$$\frac{\partial A}{\partial t} = -\frac{\delta \mathrm{CS}(A)}{\delta A} = -\star_3 F, \tag{1.22}$$

where $F = dA + A \wedge A$ is the curvature. Now a couple of miracles happen. This equation has a priori no reason to be elliptic or to have four-dimensional symmetry. But it turns out that the equation is actually a gauge-fixed version of the instanton equation $F^+ = 0$, which is elliptic modulo the gauge group and has the full four-dimensional symmetry (i.e., it is naturally defined on any oriented Riemannian four-manifold M, not necessarily of the form $W \times \mathbb{R}$ for some W). These miracles are well known to researchers on Donaldson and Floer theory, where they play an important role.

It turns out that similar miracles happen in gradient flow on the space $\mathcal{U}$ of complex-valued connections, endowed with the obvious flat Kähler metric

$$|\delta\mathcal{A}|^2 = -\int_W \mathrm{Tr}\,\delta\mathcal{A} \wedge \star\delta\overline{\mathcal{A}}. \tag{1.23}$$

This equation is a gauge-fixed version (with also the moment map set to 0, in a sense explained in [16]) of an elliptic differential equation that has full four-dimensional symmetry. This equation can be seen as a four-dimensional cousin of Hitchin's celebrated equation in two dimensions. It is an equation for a pair A, ϕ, where A is a real connection on a $G^{\vee}$-bundle $E^{\vee} \to M$, M being an oriented four-manifold, and ϕ is a one-form on M with values in $\mathrm{ad}(E^{\vee})$. The equations (for simplicity I take k real) are

$$F - \phi \wedge \phi = \star d_A\phi, \qquad d_A \star \phi = 0. \tag{1.24}$$

They can be viewed as flow equations for the complex-valued connection $\mathcal{A} = A + i\phi$ on the three-manifold W.

There is a happy coincidence: these equations, which sometimes have been called the KW equations, arise in a certain twisted version of maximally supersymmetric Yang–Mills theory ($\mathcal{N} = 4$ super Yang–Mills theory) in four dimensions [17]. We will see shortly why this relationship is relevant. For recent mathematical work on these equations, see [18, 19] and also [20].

Now we can define a Lefschetz thimble for any choice of a complex flat connection $\mathcal{A}_\rho$ on M, associated to a homomorphism $\rho : \pi_1(M) \to G_{\mathbb{C}}^{\vee}$. We work on the four-manifold $M = W \times \mathbb{R}_+$, where $\mathbb{R}_+$ is the half-line $t \geq 0$, and define the thimble Γ_ρ to consist of all complex connections $\mathcal{A} = A + i\phi$ that are boundary values (at the finite boundary of M at $W \times \{t = 0\}$) of solutions of the KW equations on M that approach $\mathcal{A}_\rho$ at infinity.

For a general M, there are various choices of ρ and some rather interesting issues that have not yet been unraveled. But now we can see what is special about knots in $\mathbb{R}^3$. Since the fundamental group of $\mathbb{R}^3$ is trivial,[9] any complex flat connection on $\mathbb{R}^3$ is equivalent to the trivial one, $\mathcal{A} = 0$. Hence there is only one Lefschetz thimble Γ_0, and any integration cycle is a multiple of this one. So, instead of integration over U to define the Jones polynomial, we can define the quantum knot invariants by integration over Γ_0:

$$Z_k(\mathbb{R}^3; K, R^{\vee}) = \frac{1}{\text{vol}} \int_{\Gamma_0} D\mathcal{A} \, \exp[ik\text{CS}(\mathcal{A})] \cdot \mathcal{W}_{R^{\vee}}(K). \qquad (1.25)$$

Here the holonomy function $\mathcal{W}_{R^{\vee}}(K)$ is viewed as a function on the Lefschetz thimble; in other words, it is evaluated for the connection $\mathcal{A} = A + i\phi$ restricted to $W \times \{0\}$ (so, in Fig. 1.6, the knot K is placed on the boundary of $W \times \mathbb{R}_+$; as in footnote 9, K does not enter the definition of the Lefschetz thimble). This definition explains why the quantum invariants of knots in $\mathbb{R}^3$ can be analytically continued away from roots of unity. Indeed, the function $\text{CS}(\mathcal{A})$ is well defined and single-valued on the Lefschetz thimble Γ_0, so there is no reason to restrict to the case that k is an integer.

The formula (1.25) for the Jones polynomial and its cousins may appear to be purely formal, but there is a reason that we can say something about it. As I have already observed, the KW equations arise in $\mathcal{N} = 4$ super Yang–Mills theory in four dimensions. This theory has a "twisted" version that localizes on the space of solutions of the KW equations. The space of all such solutions on $M = \mathbb{R}^3 \times \mathbb{R}_+$, with the requirement that $\mathcal{A} \to 0$ at ∞, is simply our Lefschetz thimble Γ_0. The upshot is that the quantum invariants of a knot in $\mathbb{R}^3$ can be computed from a path integral of $\mathcal{N} = 4$ super Yang–Mills theory in four dimensions, with a slightly subtle boundary condition [22] along the boundary at $\mathbb{R}^3 \times \{0\}$.

This is not yet obviously useful, but one more step brings us into a more accessible world, and also gives a new explanation of why the quantum knot invariants are Laurent polynomials in the variable q. The step in question was also a key step in [17] and more generally in most of the work of physicists on the supersymmetric gauge theory in

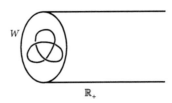

Figure 1.6 A four-manifold $M = W \times \mathbb{R}_+$, with a knot K embedded in its boundary $W \times \{0\}$.

question. This is electric–magnetic duality, the four-dimensional analog of mirror symmetry in two dimensions. $\mathcal{N} = 4$ supersymmetric Yang–Mills theory with gauge group $G^\vee$ and "coupling parameter" τ is equivalent to the same theory with $G^\vee$ replaced by its Langlands or GNO dual group, which we simply call G, and the coupling parameter $\tau^\vee$ replaced by $\tau = -1/n_\mathfrak{g} \tau^\vee$ (here $n_\mathfrak{g}$ is the ratio of the lengths squared of the long and short roots of G or equivalently of $G^\vee$).

To find a dual description in our problem, we need to ask what happens under the duality to the boundary condition at $\mathbb{R}^3 \times \{0\}$. (The analog of this question in mirror symmetry may be more familiar: what Lagrangian submanifold is mirror to a given coherent sheaf?) For the boundary condition that is related to the Lefschetz thimble, the dual boundary condition was described some years ago in [22]. It is somewhat unusual and will be described in Lecture Two. For now, I will just say that this boundary condition has the formal properties of a standard local elliptic boundary condition and has the effect of reducing to finite-dimensional spaces of solutions of the KW equations.

In the situation of Fig. 1.6, after making the duality transformation, the moduli space of solutions has expected dimension 0, and to evaluate $Z_k(\mathbb{R}^3; K, R^\vee)$, we just have to "count" (with signs, as in Donaldson theory) the number b_n of solutions for a given value n of the instanton number (for $G = SU(n)$, the instanton number is the second Chern class). The boundary conditions depend on the knot K and on the representation $R^\vee$ by which it is labeled. This is the only way that K and $R^\vee$ enter in this dual description. The path integral gives

$$Z(q; K, R^\vee) = \sum_n b_n q^n, \tag{1.26}$$

where q was defined in (1.5). This exhibits the Jones polynomial and the related quantum invariants of knots in three dimensions as "Laurent polynomials" in q with integer coefficients. I put "Laurent polynomials" in quotes because the powers of q are shifted from integers in a way that depends only on the representations, so for instance the Jones polynomial of a knot with this normalization is $q^{1/2}$ times a Laurent polynomial in q.

I have changed notation slightly to write the knot invariant as $Z(q; K, R^\vee)$ rather than $Z_k(\mathbb{R}^3; K, R^\vee)$, since this formula only works in this simple way on the three-manifold $\mathbb{R}^3$, and also in this description the natural variable is q rather than k.

Should we be surprised that duality converts a three-dimensional path integral—the analytically continued Chern–Simons path integral—into a problem of counting classical solutions? I think that the real surprise is that it is possible to arrange things so that supersymmetric localization of a four-dimensional path integral leads to a three-dimensional path integral, namely, an analytically continued version of the usual path integral of Chern–Simons gauge theory. Making this work is fairly delicate. A much more generic outcome of supersymmetric localization is a problem of counting classical solutions, and we should not be too surprised that electric–magnetic duality maps the special four-dimensional construction related to Chern–Simons theory to a more generic one that leads to counting of classical solutions. A rough analogy involves mirror symmetry in two dimensions between the A-model and the B-model. The B-model is a relatively exceptional construction in two dimensions in which supersymmetric localization leads to variation of Hodge structure rather than counting of classical solutions. Mirror symmetry converts it to the more generic[10] A-model in which one has to count classical solutions. Variation of Hodge structure is part of Hodge theory, which can be interpreted in terms of quantum mechanics or (equivalently) one-dimensional quantum field theory. (The quasi-one-dimensional nature of the B-model is what physicists mean in calling it "classical.") So mirror symmetry from the B-model to the A-model in effect maps a quantum problem in one dimension to a counting of classical solutions in two dimensions. The analog in two dimensions more is the electric–magnetic duality from a quantum problem in three dimensions—the analytically continued Chern–Simons theory—to a classical problem of counting solutions of the KW equations in four dimensions.

A final remark is that the formula (1.26), in which q appears as an instanton-counting parameter, can be viewed as a response to a challenge raised in [25], p. 299. The challenge was to find a description of the Jones polynomial in which q would be associated to instanton number (the integral of the second Chern class) in four dimensions.

1.2 Lecture Two

As we discussed in the last lecture, quantum knot invariants of a simple Lie group $G^\vee$ on a three-manifold W can be computed by counting solutions of a certain system of elliptic partial differential equations, with gauge group the dual group G, on the four-manifold $M = W \times \mathbb{R}_+$. The equations are the KW equations

$$F - \phi \wedge \phi = \star d_A \phi, \qquad d_A \star \phi = 0, \tag{1.27}$$

where A is a connection on a G-bundle $E \to M$ and $\phi \in \Omega^1(M, \mathrm{ad}(E))$. The boundary conditions at the finite end of $M = W \times \mathbb{R}_+$ depend on the knot, as indicated in Fig. 1.6. The boundary conditions at the infinite end of M say that $\mathcal{A} = A + i\phi$ must approach a complex-valued flat connection. Exactly what we have to do depends on what we want to get, but in one very important case there is a simple answer. For $W = \mathbb{R}^3$, meaning that we are studying knots in $\mathbb{R}^3$, a flat connection is gauge-equivalent to zero and we require that $\mathcal{A} \to 0$ at infinity, in other words $A, \phi \to 0$. In this lecture, we are only going to discuss the case of $W = \mathbb{R}^3$.

For $W = \mathbb{R}^3$, the difference between $G^\vee$ and G is going to be important primarily when they have different Lie algebras, since for instance there is no second Stiefel–Whitney class to distinguish SO(3) from SU(2). So the difference will be most important if $G^\vee = \mathrm{SO}(2n + 1)$ and $G = \mathrm{Sp}(2n + 1)$, or vice versa. In fact, we will see later that something very interesting happens precisely for $G^\vee = \mathrm{SO}(2n + 1)$ (or its double cover Spin$(2n + 1)$).

To compute quantum knot invariants, we are supposed to "count" the solutions of the KW equations with fixed instanton number. The instanton number is defined as

$$P = \frac{1}{8\pi^2} \int_M \mathrm{Tr}\, F \wedge F, \tag{1.28}$$

where the trace is an invariant quadratic form defined so that (for simply connected G on a compact four-manifold M without boundary), P takes integer values. For $G = SU(n)$, we can take Tr to be the trace in the n-dimensional representation; then P is the second Chern class. Just as in Donaldson theory, the "count" of solutions is made with signs. The sign with which a given solution contributes is the sign of the determinant of the linear elliptic operator that arises by linearizing the KW equations about a given solution. (For physicists, this is the fermion determinant.)

Let b_n be the "number" of solutions of instanton number $P = n$. One forms the series

$$Z(q) = \sum_n b_n q^n. \qquad (1.29)$$

One expects that b_n vanishes for all but finitely many n. Given this, $Z(q)$ (which depends on the knot K and a representation $R^\vee$, though we now omit these in the notation) is a Laurent polynomial in q (times q^c for some fixed $c \in \mathbb{Q}$, as explained shortly). For example, if $G^\vee = SU(2)$ and the knot is labeled by the two-dimensional representation, then $Z(q)$ is expected to be the Jones polynomial.

In all of this, the knot and representation are encoded entirely in the boundary condition at the finite end of M, as sketched in Fig. 1.6. The instanton number P is an integer if M is compact and without boundary, but we are not in that situation. To make P into a topological invariant, we need a trivialization of the bundle E at both the finite and infinite ends of M. The trivialization at the infinite end comes from the requirement that $A, \phi \to 0$ at infinity. The trivialization at the finite end depends on the boundary condition, which I have not yet described. With this boundary condition, P is offset from being integer-valued by a constant that depends only on the knots K_i in W and the representations $R_i^\vee$ labeling them. This is why $Z(q)$ is not quite a Laurent polynomial in q, but is q^c times such a Laurent polynomial, where c is completely determined by the representations $R_i^\vee$ and the framings of the K_i.

Given this description of the Jones polynomial and related knot invariants, I want to explain how to associate these knot invariants with a homology theory (which is expected to coincide with Khovanov homology). I should say that the original work by physicists associating vector spaces to knots was by Ooguri and Vafa [26] (following earlier work associating vector spaces to homology cycles in a Calabi–Yau manifold [27, 28]). After the invention of Khovanov homology of knots [29], a relation of the Ooguri–Vafa construction to Khovanov homology was proposed [30]. What I will be summarizing here is a parallel construction [21] in gauge theory language. The arguments are probably more self-contained, though it is hard to make this entirely clear in these lectures; the construction is more uniform for all groups and representations; and I believe that the output is something that mathematicians will be able to grapple with even without a full understanding of the underlying quantum field theory. I should also say that my proposal for Khovanov homology is qualitatively similar to ideas by Seidel and Smith and by Kronheimer and Mrowka, and probably others, and is expected to be a mirror to a construction by Cautis and Kamnitzer. See [31–35] for references to this mathematical work.

Let S be the set of solutions of the KW equations. (It is expected that for a generic embedding of a knot or link in $\mathbb{R}^3$, the KW equations have only finitely many solutions and these are non-degenerate: the linearized operator has trivial kernel and cokernel.) We define a vector space $\mathcal{V}$ by declaring that for every $i \in S$, there is a corresponding basis vector $|i\rangle$. On $\mathcal{V}$, we will have two "conserved quantum numbers," which will be the instanton number P and a second quantity that I will call the "fermion number" F. I have already explained that P takes values in $\mathbb{Z} + c$, where c is a fixed constant that depends only on the choices of representations and framings. F (which will be defined as a certain Morse index) will be integer-valued. The states $|i\rangle$ corresponding to solutions will be eigenstates of F. We consider the state $|i\rangle$ to be "bosonic" or "fermionic" depending on whether it has an even or odd value of F. So the operator distinguishing bosons from fermions is $(-1)^F$. F will be defined so that if the solution i contributed $+1$ to the counting of KW solutions, then $|i\rangle$ has even F, and if it contributed -1, then $|i\rangle$ has odd F.

Let us see how we would rewrite in this language the quantum knot invariant

$$Z(q) = \sum_n b_n q^n. \tag{1.30}$$

Here a solution $i \in S$ with instanton number n_i and fermion number f_i contributes $(-1)^{f_i}$ to b_{n_i}, so it contributes $(-1)^{f_i} q^{n_i}$ to $Z(q)$. So an equivalent formula is

$$Z(q) = \sum_{i \in S} (-1)^{f_i} q^{n_i} = \mathrm{Tr}_{\mathcal{V}} (-1)^F q^P. \tag{1.31}$$

So far, we have not really done anything except to shift things around. However, on $\mathcal{V}$, we will also have a "differential" Q, which is an operator that commutes with P but increases F by 1, and obeys $Q^2 = 0$. These statements mean that we can define the *cohomology* of Q, which we denote by $\mathcal{H}$, and moreover that $\mathcal{H}$ is $\mathbb{Z} \times \mathbb{Z}$-graded, with the two gradings determined by P and F (we simplify slightly, ignoring the fact that the eigenvalues of P are really in a coset $\mathbb{Z} + c$).

The importance of passing from $\mathcal{V}$ to $\mathcal{H}$ is that $\mathcal{H}$ is a topological invariant, while $\mathcal{V}$ is not. If one deforms a knot embedded in $\mathbb{R}^3$, then solutions of the KW equations on $M = \mathbb{R}^3 \times \mathbb{R}_+$ will appear and disappear, so $\mathcal{V}$ will change. But $\mathcal{H}$ does not change. This $\mathcal{H}$ is the candidate for the Khovanov homology.

Instead of defining $Z(q)$ as a trace in $\mathcal{V}$ via (1.31), we can define it as a trace in $\mathcal{H}$:

$$Z(q) = \mathrm{Tr}_{\mathcal{H}} (-1)^F q^P. \tag{1.32}$$

So, here, $Z(q)$ is expressed as an "Euler characteristic," i.e., as a trace in which bosonic and fermionic states cancel, in the invariantly defined cohomology $\mathcal{H}$. The reason that we can equally well write $Z(q)$ as a trace in $\mathcal{V}$ or in $\mathcal{H}$ is standard: the difference between $\mathcal{V}$ and $\mathcal{H}$ is that in passing from $\mathcal{V}$ to $\mathcal{H}$, pairs of states disappear that make vanishing contributions to $Z(q)$. (Such a pair consists of a bosonic state and a fermionic state with the same value of P and values of F differing by 1.)

Defining the $\mathbb{Z} \times \mathbb{Z}$-graded vector space $\mathcal{H}$ and not just the trace $Z(q)$ adds information for two reasons. One reason is simply that the fermion number F is really $\mathbb{Z}$-valued, and this is part of the $\mathbb{Z} \times \mathbb{Z}$ grading of $\mathcal{H}$. When we pass from $\mathcal{H}$ to the trace $Z(q)$, we only remember F modulo 2, and here we lose some information. The second reason is that one can define natural operators acting on $\mathcal{H}$, and how they act adds more information. For it to make sense to define the action of operators, we need a "quantum Hilbert space" $\mathcal{H}$ for them to act on, and not just a function $Z(q)$. I explain later how to define natural operators, associated to link cobordisms, that act on $\mathcal{H}$.

The ability to do all this rests on the following facts about the KW equations. (These facts were also discovered by A. Haydys [36].) I will just state these facts as facts—which one can verify by a short calculation—without describing the quantum field theory construction that motivated me to look for them [21]. We consider the KW equations on a four-manifold $M = W \times \mathcal{I}$, where W is a three-manifold with local coordinates x^i, $i = 1, 2, 3$, and $\mathcal{I}$ is a one-manifold parametrized by y. (In our application, $\mathcal{I}$ will be $\mathbb{R}_+$. In the first lecture, $\mathbb{R}_+$ was introduced as the direction of a gradient flow, and parametrized by "time," but now we interpret $\mathcal{I}$ as a "space" direction; we are about to introduce a new "time" direction.) We write $\phi = \sum_{i=1}^3 \phi_i \mathrm{d}x^i + \phi_y \mathrm{d}y$. Now we replace M by a five-manifold $X = \mathbb{R} \times M$, where $\mathbb{R}$ is parametrized by a new "time" coordinate t. We convert the four-dimensional KW equations on M into five-dimensional equations on X by simply replacing ϕ_y, wherever it appears, by a covariant derivative in the new time direction:

$$\phi_y \to \frac{D}{Dt} = \frac{\partial}{\partial t} + [A_t, \cdot]. \tag{1.33}$$

If we make this substitution in a random differential equation containing ϕ_y, we will not get a differential equation but a differential operator. In the case of the KW equations, ϕ_y appears only inside commutators $[\phi_i, \phi_y]$ and covariant derivatives $D_\mu \phi_y$, and the substitutions proceed by

$$[\phi_i, \phi_y] \to -D_t \phi_i, \qquad D_\mu \phi_y \to [D_\mu, D_t] = F_{\mu t}. \tag{1.34}$$

This is enough to show that the substitution does give a differential equation. Generically, the differential equation obtained this way would not be well posed, where here "well posed" means "elliptic." Essential to make our story work is that the five-dimensional equation obtained in the case of the KW equations from the substitution $\phi_y \to D/Dt$ actually is elliptic. This is not hard to verify if one suspects it.

The five-dimensional equation has a four-dimensional symmetry that is not obvious from what we have said so far. We started on $M = W \times \mathcal{I}$, with W a three-manifold, and then, via $\phi_y \to D/Dt$, we replaced M with $X = \mathbb{R} \times M = \mathbb{R} \times W \times \mathcal{I}$. It turns out that here $\mathbb{R} \times W$ can be replaced by any oriented four-manifold Z, and our equation can be naturally defined on[11] $X = Z \times \mathcal{I}$. At a certain point, we will make use of this four-dimensional symmetry.

Another essential fact is that the five-dimensional equation that we get this way can be formulated as a gradient flow equation

$$\frac{d\Phi}{dt} = -\frac{\delta\Gamma}{\delta\Phi}, \tag{1.35}$$

for a certain functional $\Gamma(\Phi)$ (here all fields A, ϕ are schematically combined into Φ). This means that we are in the situation explored by Floer when he defined Floer cohomology in the 1980s: modulo analytic subtleties, we can define an infinite-dimensional version of Morse theory, with Γ as a middle-dimensional Morse function.

In Morse homology (i.e., in Morse theory formulated by counting of gradient flow lines [37, 38]), one defines a vector space $\mathcal{V}$ with a basis vector $|i\rangle$ for every critical point of Γ. $\mathcal{V}$ is $\mathbb{Z}$-graded by a "fermion number" operator F that assigns to $|i\rangle$ the Morse index f_i of the critical point $i \in \mathcal{S}$. If, as in our case, the Morse function is defined on a space that has connected components labeled by another quantity P (in our case, P is the instanton number operator), then $\mathcal{V}$ is also graded by the value of P for a given critical point. On $\mathcal{V}$, one defines a "differential" $Q : \mathcal{V} \to \mathcal{V}$ by

$$Q|i\rangle = \sum_{j\in\mathcal{S}|f_j=f_i+1} n_{ij}|j\rangle, \tag{1.36}$$

where for each pair of critical points i, j with $f_j = f_i + 1$, we define n_{ij} as the "number" of solutions of the gradient flow equation

$$\frac{d\Phi}{dt} = -\frac{\partial\Gamma}{\partial\Phi}, \qquad -\infty < t < \infty \tag{1.37}$$

that start at i in the far past and end at j in the far future. In the counting, one factors out by the time-translation symmetry, and one includes a sign given by the sign of the fermion determinant, that is, the sign of the determinant of the linear operator obtained by linearizing the gradient flow equation. On a finite-dimensional compact manifold B, the cohomology of Q is simply the cohomology of B with integer coefficients. Floer's basic idea (which can be interpreted physically as a generalization of the procedure just described to quantum field theories in higher dimension) is that the cohomology of Q makes sense in an infinite-dimensional setting provided the flow equation is elliptic and certain compactness properties hold. (In our present context, the flow equation is certainly elliptic, but the necessary compactness properties have not yet been proved.)

When we follow this recipe in the present context, the time-independent solutions in five dimensions are just the solutions of the KW equations in four dimensions (with A_t reinterpreted as ϕ_y), since when we ask for a solution to be time-independent, we undo what we did to go from four to five dimensions. So the space $\mathcal{V}$ on which the differential of Morse theory acts is the same space we introduced before in writing the quantum knot invariant $Z(q)$ as a trace. Moreover, in our application, the procedure described in the

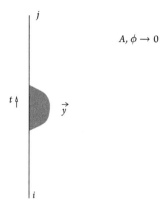

Figure 1.7 Matrix elements of the differential Q are computed by counting solutions of the five-dimensional equations interpolating between two given four-dimensional solutions i in the past and j in the future. The solutions vanish for $y \to \infty$, and at $y = 0$ obey boundary conditions that will be described in the text.

last paragraph means that the matrix elements n_{ij} of the differential should be computed by "counting" the five-dimensional solutions that interpolate from a KW solution i in the past to a KW solution j in the future (Fig. 1.7).

A conspicuous gap here is that I have not yet described the boundary condition that should be used at the finite end of X, in other words at $y = 0$. Before doing so, I want to describe something interesting that happens in Khovanov homology for certain gauge groups. In the study of Khovanov homology for $G^\vee = \mathrm{SU}(2)$, it has been found [39] that there are two variants of the theory, called "even" and "odd" Khovanov homology. They are defined using a complex $\mathcal{V}$ that additively is the same in the two cases, but on this complex one defines two different differentials, say Q_+ for the even theory and Q_- for the odd theory. They are both defined over $\mathbb{Z}$ and they are congruent mod 2, so their cohomologies, which are called even and odd Khovanov homology, are isomorphic if one reduces mod 2. Why would this happen in our framework and for what groups should we expect it to happen? I claim that we should use the exceptional isomorphism $\mathrm{SU}(2) \cong \mathrm{Spin}(3)$ and that in general the bifurcation into even and odd Khovanov homology will occur precisely for $G^\vee = \mathrm{Spin}(2n + 1), n = 1, 2, 3, \ldots$.

In general, the cohomology of a manifold B can be twisted by a flat complex line bundle $\mathcal{L}$. Instead of the ordinary cohomology $H^i(B, \mathbb{Z})$, we can consider the cohomology with values in $\mathcal{L}$, $H^i(B, \mathcal{L})$. There is a Morse theory recipe to compute this, by slightly modifying the procedure described above. The possible $\mathcal{L}$'s are classified by $\mathrm{Hom}(\pi_1(B), \mathbb{C}^*)$. In the present context, B is a function space, consisting of pairs A, ϕ on $M = W \times \mathbb{R}_+$ (which represent initial data for solutions on $X = \mathbb{R} \times M$, where $\mathbb{R}$ is parametrized by "time," and which obey certain boundary conditions). We only care about the pairs A, ϕ up to G-valued gauge transformations (which, because of the boundary conditions, are trivial on the finite and infinite boundaries of M). For $W = \mathbb{R}^3$, this means that $\pi_1(B) = \pi_4(G)$. For the simple Lie groups, we have

$$\pi_4(G) = \begin{cases} \mathbb{Z}_2 & \text{for } G = \mathrm{Sp}(2n) \text{ or } \mathrm{Sp}(2n)/\mathbb{Z}_2, \quad n \geq 1, \\ 0 & \text{otherwise.} \end{cases}$$

So Khovanov homology is unique unless $G^\vee = \mathrm{SO}(2n + 1)$, $G = \mathrm{Sp}(2n)$ (or $G^\vee = \mathrm{Spin}(2n + 1)$, $G = \mathrm{Sp}(2n)/\mathbb{Z}_2$), for some $n \geq 1$, in which case there are two versions of Khovanov homology. Concretely, an $\mathrm{Sp}(2n)$ bundle on a five-dimensional spin manifold Y (with a trivialization at infinity along Y) has a $\mathbb{Z}_2$-valued invariant ζ derived from $\pi_4(\mathrm{Sp}(2n)) = \mathbb{Z}_2$. ($\zeta$ is defined as the mod 2 index of the Dirac operator valued in the fundamental representation of $\mathrm{Sp}(2n)$.) When we define the differential by counting five-dimensional solutions, we have the option to modify the differential by including a factor of $(-1)^\zeta$. This gives a second differential Q' that still obeys $(Q')^2 = 0$, and is congruent mod 2 to the differential Q that is obtained without the factor of $(-1)^\zeta$. The two theories associated to Q and Q' are the candidates for the two versions of Khovanov homology.[12]

Next I come to an explanation of the boundary conditions that we impose on the four- or five-dimensional equations. These boundary conditions are crucial, since for instance it is only via the boundary conditions that knots enter. First I will describe the boundary condition in the absence of knots. It is essentially enough to describe the boundary condition in four dimensions rather than five (once one understands it, the lift to five dimensions is fairly obvious), and as the boundary condition is local, we assume initially that the boundary of the four-manifold is just $\mathbb{R}^3$. So we work on $M = \mathbb{R}^3 \times \mathbb{R}_+$. (This special case is anyway the right case for the Jones polynomial, which concerns knots in $\mathbb{R}^3$ or equivalently S^3.)

As a preliminary to describing the boundary condition, I need to describe an important equation in gauge theory, which is Nahm's equation. This is a system of ordinary differential equations for a triple X_1, X_2, X_3 of elements of $\mathfrak{g}$, the Lie algebra of a compact Lie group G. The equations read

$$\frac{\mathrm{d}X_1}{\mathrm{d}y} + [X_2, X_3] = 0, \tag{1.38}$$

and cyclic permutations thereof. On an open half-line $y > 0$, Nahm's equations have the special solution

$$X_i = \frac{t_i}{y}, \tag{1.39}$$

where t_i are elements of $\mathfrak{g}$ that obey the $\mathfrak{su}(2)$ commutation relations $[t_1, t_2] = t_3$ and cyclic permutations. Thus the t_i are the images of a standard basis of $\mathfrak{su}(2)$ under a homomorphism $\varrho : \mathfrak{su}(2) \to \mathfrak{g}$. This singular solution of Nahm's equations has been important in numerous applications, in work by Nahm, Kronheimer, Atiyah and Bielawski, and others [42–45]. We will use it to define an elliptic boundary condition on the KW equations and their five-dimensional cousins.

In fact, Nahm's equations can be embedded in the KW equations (1.27) on $\mathbb{R}^3 \times \mathbb{R}_+$. We look for a solution that (i) is invariant under translations of $\mathbb{R}^3$, (ii) has the property that $A = 0$ and (iii) has the property that $\phi = \sum_{i=1}^{3} \phi_i \, dx^i + 0 \cdot dy$. For solutions satisfying these conditions, the KW equations reduce to Nahm's equations

$$\frac{d\phi_1}{dy} + [\phi_2, \phi_3] = 0, \tag{1.40}$$

and cyclic permutations. So the "Nahm pole" gives a special solution

$$\phi_i = \frac{t_i}{y}. \tag{1.41}$$

We define a boundary condition by saying that we only allow solutions that are asymptotic to this one for $y \to 0$, modulo less singular terms. See [46] for a proof that, for any ϱ, this boundary condition is elliptic and has regularity properties similar to those of more standard elliptic boundary conditions such as Dirichlet or Neumann. (Actually, in that paper, a somewhat wider class of Nahm pole boundary conditions is analyzed.) For applications to the Jones polynomial and Khovanov homology, we take $\varrho : \mathfrak{su}(2) \to \mathfrak{g}$ to be a principal embedding in the sense of Kostant (for example, for $G = SU(n)$, this means that the n-dimensional representation of G transforms as an irreducible representation of $\mathfrak{su}(2)$).

This is the boundary condition that we want at $y = 0$, in the absence of knots. For the simplest application to the quantum knot invariants and Khovanov homology, we require that $A, \phi \to 0$ for $y \to \infty$. With these conditions at $y = 0, \infty$, and suitable conditions for $x^i \to \infty$, it is possible to prove [46] that the Nahm pole solution is the only solution on $\mathbb{R}^3 \times \mathbb{R}_+$. This corresponds to the statement that the Khovanov homology of the empty link is of rank 1.

Now I should explain how the boundary condition is modified along a knot K. This will be done by requiring a more subtle singularity along the knot. The local model is that the boundary of M is $\mathbb{R}^3$ and the knot K is a copy of $\mathbb{R} \subset \mathbb{R}^3$. The boundary condition is defined by giving a model solution on $\mathbb{R}^3 \times \mathbb{R}_+$ that away from K has the now familiar Nahm pole singularity at $y = 0$, but has a more complicated singular behavior along K. The model solution is invariant under translations along K, so it can be obtained by solving some reduced equations on $\mathbb{R}^2 \times \mathbb{R}_+$. In the reduced picture, K corresponds to a point in $\mathbb{R}^2 \times \{y = 0\}$ (Fig. 1.8). The model solution depends on the choice of a representation $R^\vee$ of the dual group $G^\vee$. Near a boundary point disjoint from K, the model solution has the usual Nahm pole singularity, but near K it has a more complicated singularity. The relevant model solutions can be found in closed form. This has been done in [21], Section 3.6, for $G^\vee = SU(2)$, and in [47] for any $G^\vee$. I will not describe these reduced solutions here, except to say that the singularity along K is a more complicated cousin of the singularity used in [17] to describe the geometric Hecke transformations of the geometric Langlands correspondence.

The model solution has a singularity that, along the boundary, is of codimension 2. When we go to five dimensions, the singularity remains of codimension 2, so now,

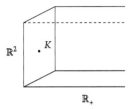

Figure 1.8 The boundary condition in the presence of a knot K is determined by a model solution of a reduced equation on $\mathbb{R}^2 \times \mathbb{R}_+$. Near a generic boundary point, the model solution has the Nahm pole singularity, but it has a more complicated singular behavior near the boundary point corresponding to K. This model solution depends on the choice of a representation $R^\vee$ of the dual group $G^\vee$.

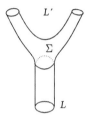

Figure 1.9 A two-dimensional surface Σ that represents a "link cobordism" from a link L in the past to another link L' in the future. In the example shown, for simplicity, L is an unknot and L' consists of two unlinked unknots. Σ is embedded in the boundary of $X = \mathbb{R}^4 \times \mathbb{R}_+$.

as the boundary is a four-manifold, the singularity is supported on a two-dimensional surface Σ, not a knot (Fig. 1.9). A boundary condition modified on a two-surface in the boundary is what we need to define the "morphism" of Khovanov homology associated to a "link cobordism." In other words, given a two-surface Σ that interpolates between one link L in the past and another link L' in the future, as sketched in the figure, counting solutions with boundary conditions modified along Σ gives the matrix element for a time-dependent transition from a physical state (a cohomology class of Q) in the past in the presence of L to a physical state in the future in the presence of L'. The morphisms that are defined this way have the formal properties that are expected in Khovanov homology.

There is another reason that it is important to describe the reduced equations in three dimensions. To compute the Jones polynomial, we need to count certain solutions in four dimensions; knowledge of these solutions is also the first step in constructing the candidate for Khovanov homology. How are we supposed to describe four-dimensional solutions? A standard strategy, often used in Floer theory and its cousins, involves "stretching" the knot in one direction, in the hope of reducing to a piecewise description by solutions in one dimension less (Fig. 1.10). To get anywhere with such an analysis, we need to be able to solve the three-dimensional reduced equations.

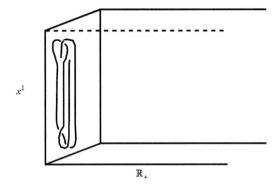

Figure 1.10 Stretching a knot in one dimension, to reduce to a description in one dimension less. One of the directions in $\mathbb{R}^3$—here labeled as x^1—plays the role of "time." After stretching of the knot, one hopes that a solution of the equations becomes almost everywhere nearly independent of x^1. If so, a knowledge of the solutions that are actually independent of x^1 can be a starting point for understanding four-dimensional solutions. This type of analysis will fail at the critical points of the function x^1 along a knot or link; a correction has to be made at those points.

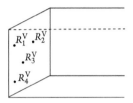

Figure 1.11 $\mathbb{R}^2 \times \mathbb{R}_+$ with n boundary points labeled by representations $R_1^\vee, \ldots, R_n^\vee$ of the dual group, in this case with $n = 4$.

Another way to make the point is as follows. Most mathematical definitions of Khovanov homology proceed, directly or implicitly, by defining a category of objects associated to a two-sphere (or, in some versions, a copy of $\mathbb{C} = \mathbb{R}^2$) with marked points that are suitably labeled. In the present approach, this category should be a category associated to solutions of the reduced three-dimensional equations on $\mathbb{R}^2 \times \mathbb{R}_+$, with finitely many points in $\mathbb{R}^2 \times \{y = 0\}$ labeled by representations of the dual group $G^\vee$ (Fig. 1.11). By analyzing this situation, Gaiotto and I [48] were able to get a fairly clear framework for understanding the relevant category. (We found that to make this program work nicely, we had to perturb to a slightly more generic version of the problem than I have described.) For $G^\vee = \mathrm{SU}(2)$, the category is expected to be a Fukaya–Seidel category (an A-model category with a superpotential) where the target space is a moduli space of monopoles on $\mathbb{R}^3$, and the superpotential encodes the positions of the knots. This category has not yet been analyzed in any detail, but for the more modest problem of understanding the Jones polynomial (rather than Khovanov homology), we were able to get a reasonably satisfactory picture.

In doing this, we used a simplification that can be achieved by modifying the condition on how a solution of the KW equations should behave for $y \to \infty$. We kept the condition that $A \to 0$ at infinity, but required ϕ to approach $\sum_{i=1}^{3} c_i \, dx^i$, where the c_i are a prescribed triple of commuting elements of $\mathfrak{g}$. (It is known [44] that this is a convenient asymptotic condition in the study of Nahm's equations.) The counting of solutions of an elliptic differential equation is invariant under continuous deformations, as long as certain conditions are satisfied. So it is expected that the counting of solutions of the KW equations is independent of $\vec{c} = (c_1, c_2, c_3)$. By exploiting a suitable choice of $\vec{c}$, we were able to relate the counting of solutions of the KW equations to the vertex model for the Jones polynomial, which was the starting point in Lecture One. Notice that $\vec{c}$ defines a direction in $\mathbb{R}^3$. This direction determines the knot projection that is used in the vertex model.

Recent work reconsidering the Fukaya–Seidel category from a physical point of view [49] may be helpful in understanding better the categories that arise in the present context when a knot is stretched in one dimension.

Acknowledgments

This chapter is based on lectures presented at the Clay Research Conference at Oxford University, and also at the Galileo Galilei Institute in Florence, the University of Milan, Harvard University, and the University of Pennsylvania.

This research was supported in part by NSF Grant PHY-1314311.

Notes

1. A framing of a knot in $\mathbb{R}^3$ is a trivialization of the normal bundle to the knot. If a knot is given with a projection to $\mathbb{R}^2$, then the normal direction to $\mathbb{R}^2$ gives a framing. A change of framing multiplies the sum that comes from the vertex model by an integer power of $q^{3/4}$. A knot in $\mathbb{R}^3$ can be given a canonical framing, and therefore the Jones polynomial can be expressed as an invariant of a knot without a choice of framing. But this is not always convenient.

2. $G^\vee$ is a common notation for the Langlands or GNO (Goddard–Nuyts–Olive) dual of a simple Lie group G. Duality will later enter our story, and we will have two descriptions involving a dual pair of groups G and $G^\vee$. We write $G^\vee$ for the group that is used in the conformal field theory and Chern–Simons descriptions, because this will agree better with the usual terminology concerning the Langlands correspondence. Similarly, we write $R^\vee$ for a representation of $G^\vee$ and $E^\vee$ for a $G^\vee$ bundle.

3. If $G^\vee$ is connected and simply connected, then inevitably any $G^\vee$-bundle over a three-manifold is trivial.

4. The analog of this for any simple $G^\vee$ is $q = \exp[2\pi i / n_\mathfrak{g}(k + h^\vee)]$, where $h^\vee$ is the dual Coxeter number of $G^\vee$ and $n_\mathfrak{g}$ is the ratio of the lengths squared of the long and short roots of $G^\vee$. In the dual description that we come to later, q is always the instanton-counting parameter.

5. An analytic continuation can be made away from integer k, using ideas we explain later. On a generic three-manifold, the continued function has an essential singularity at $k = \infty$ with Stokes phenomena. It is not a function of q.

6. I am not even entirely convinced that it is true. What was found in [13] is that the volume conjecture for a certain knot is valid if and only if a certain invariant of that knot is non-zero. (This invariant is the coefficient of a thimble associated to a flat $G_{\mathbb{C}}^{\vee} = \mathrm{SL}(2, \mathbb{C})$ connection of maximal volume when the standard real integration cycle U is expressed in terms of Lefschetz thimbles, along the lines of (1.20).) It is not clear why this invariant is non-zero for all knots.

7. If instead we take $k \to \infty$ with fixed n, we do not include $W_{R^{\vee}}(K)$ in the definition of a critical point; we simply view it as a function that can be evaluated at a critical point of $\mathrm{CS}(\mathcal{A})$. We will follow this second procedure later.

8. For a more detailed explanation of the following, see [16].

9. As in footnote 7, there are two fruitful approaches to the present subject, which differ by whether, in getting a semiclassical limit, we (i) take $k \to \infty$ with fixed n/k, or (ii) take $k \to \infty$ with fixed n. In approach (i), a complex critical point is a complex flat connection on the knot complement $W \backslash K$. We followed this approach in our discussion of the volume conjecture. In approach (ii), which is more convenient here and in Lecture Two, a complex critical point is simply a complex flat connection on W. In approach (ii), the holonomy function $\mathcal{W}_{R^{\vee}}(K)$ does not enter the definition of a Lefschetz thimble, but is just one factor in the function that we want to integrate over the thimble. The two viewpoints are described more thoroughly in [21].

10. Generalized complex geometry, in which a family of A-models can degenerate to a B-model [23], gives a precise framework for the statement that the A-model is more generic than the B-model. For one specific illustration of this, let X be a hyper-Kahler manifold, which comes with a $\mathbb{CP}^1$ family of complex structures. A generalized complex structure on X can be associated to any pair $J_+, J_- \in \mathbb{CP}^1$. The associated two-dimensional topological field theory is a B-model if and only if $J_+ = J_-$, and otherwise is equivalent to an A-model. So, in this context, the A-model is more generic. See [24] for more.

11. More generally [36], the equation can be defined on any oriented five-manifold X endowed with an everywhere-non-zero vector field (which for $X = Z \times \mathcal{I}$ we take to be $\partial/\partial y$).

12. Odd Khovanov homology of $\mathrm{SU}(2)$ is known [40] to be related to the supergroup $\mathrm{OSp}(1|2)$. This connection will be explored elsewhere from a physical point of view [41].

References

[1] V. F. R. Jones, *A polynomial invariant for links via von Neumann algebras*, Bull. AMS **12** (1985) 103–111.

[2] V. F. R. Jones, *On knot invariants related to some statistical mechanical models*, Pacific J. Math. **137** (1989) 311–334.

[3] L. H. Kauffman, *Knots And Physics* (World Scientific, 1991).

[4] A. Tsuchiya and Y. Kanie, *Vertex operators in conformal field theory on $\mathbb{P}^1$ and monodromy representations of the braid group*, Adv. Stud. Pure Math. **16** (1987) 297–372.

[5] E. Witten, *Quantum field theory and the Jones polynomial*, Commun. Math. Phys. **121** (1989) 351–399.

[6] J. Fröhlich and C. King, *The Chern–Simons theory and knot polynomials*, Commun. Math. Phys. **126** (1989) 167–199,

[7] R. M. Kashaev, *The hyperbolic volume of knots from the quantum dilogarithm*, Lett. Math. Phys. **39** (1997) 269–275.

[8] H. Murakami and J. Murakami, *The colored Jones polynomial and the simplicial volume of a knot*, Acta Math. **186** (2001) 85–104.

[9] H. Murakami, J. Murakami, M. Okamoto, T. Takata, and Y. Yokota, *Kashaev's conjecture and the Chern–Simons invariants of knots and links*, Experiment. Math. **11** (2002) 427–435.

[10] R. M. Kashaev and O. Tirkkonen, *A proof of the volume conjecture for torus knots*, J. Math. Sci. (NY) **115** (2003) 2033–2036; arXiv:math/9912210 [math.GT].

[11] S. Gukov, *Three-dimensional quantum gravity, Chern–Simons theory, and the A-polynomial*, Commun. Math. Phys. **255** (2005) 577–627.

[12] H. Murakami, *Asymptotic behaviors of the colored Jones polynomials of a torus knot*, Int. J. Math. **15** (2004) 547–555.

[13] E. Witten, *Analytic continuation of the Jones polynomial*, arXiv:1001.2933 [hep-th].

[14] D. Freed and R. Gompf, *Computer calculations of Witten's 3-manifold invariant*, Commun. Math. Phys. **141** (1991) 79–117.

[15] L. Jeffrey, *Chern–Simons–Witten invariants of lens spaces and torus bundles and the semi-classical approximation*, Commun. Math. Phys. **147** (1992) 563–604.

[16] E. Witten, *Khovanov homology and gauge theory*, in R. Kirby, V. Krushkal and Z. Wang, eds., *Proceedings of the FreedmanFest, Berkeley, 2011* (Mathematical Sciences Publishers, 2012), pp. 291–308; arXiv:1108.3103 [math.GT].

[17] A. Kapustin and E. Witten, *Electric–magnetic duality and the geometric Langlands program*, Commun. Numb. Theory Phys. **1** (2007) 1–236; hep-th/0604151.

[18] C. H. Taubes, *Compactness theorems for SL(2; ℂ) generalizations of the anti-self dual equations, Part I*, arXiv:1307.6447 [math.DG].

[19] C. H. Taubes, *The zero loci of ℤ/2 harmonic spinors in dimension 2, 3 and 4*, arXiv:1407.6206 [math.DG].

[20] M. Gagliardo and K. Uhlenbeck, *The geometry of the Kapustin–Witten equations*, J. Fixed Point Theory Appl. **11** (2012) 185–198.

[21] E. Witten, *Fivebranes and knots*, Quantum Topol. **3** (2012) 1–137; arXiv:1101.3216 [hep-th].

[22] D. Gaiotto and E. Witten, *Supersymmetric boundary conditions in $\mathcal{N}=4$ super Yang–Mills theory*, J. Stat. Phys. **135** (2009) 789–855; arXiv:0804.2902 [hep-th].

[23] N. Hitchin, *Generalized Calabi–Yau manifolds*, Q. J. Math. **54** (2003) 281–308; arXiv:math/0209099 [math.DG].

[24] M. Gualtieri, *Generalized complex geometry*, arXiv:math/0401221 [math.DG].

[25] M. F. Atiyah, *New invariants of three and four dimensional manifolds*, in *The Mathematical Heritage Of Herman Weyl*, Proc. Symp. Pure Math. **48** (1988) 285–300.

[26] H. Ooguri and C. Vafa, *Knot invariants and topological strings*, Nucl. Phys. B **577** (2000) 419–438; arXiv:hep-th/9912123.

[27] R. Gopakumar and C. Vafa, *On the gauge theory/geometry correspondence*, Adv. Theor. Math. Phys. **3** (1999) 1415–1443; arXiv:hep-th/9811131.

[28] R. Gopakumar and C. Vafa, *M-theory and topological strings—I, II*, arXiv: hep-th/9809187, hep-th/9812127.

[29] M. Khovanov, *A categorification of the Jones polynomial*, Duke. Math. J. **101** (2000) 359–426.

[30] S. Gukov, A. S. Schwarz and C. Vafa, *Khovanov–Rozansky homology and topological strings*, Lett. Math. Phys. **74** (2005) 53–74; arXiv: hep-th/0412243.

[31] S. Cautis and J. Kamnitzer, *Knot homology via derived categories of coherent sheaves I*, $\mathfrak{sl}(2)$ *case*, arXiv:math/0701194 [math.AG].

[32] P. Seidel and I. Smith, *A link invariant from the symplectic geometry of nilpotent slices*, Duke Math. J. **134** (2006) 453–514; arXiv:math/0405089 [math.SG].

[33] J. Kamnitzer, *The Beilinson–Drinfeld Grassmannian and symplectic knot homology*, arXiv:0811.1730 [math.QA].

[34] P. B. Kronheimer and T. S. Mrowka, *Knot homology groups from instantons*, arXiv:0806.1053 [math.GT].

[35] P. B. Kronheimer and T. S. Mrowka, *Khovanov homology is an unknot-detector*, arXiv:1005.4346 [math.GT].

[36] A. Haydys, *Fukaya–Seidel category and gauge theory*, Symplectic Geom. **13** (2015) 151–207; arXiv:1010.2353 [math.SG].

[37] E. Witten, *Supersymmetry and Morse theory*, J. Diff. Geom. **17** 661 (1982).

[38] M. Hutchings, *Lecture Notes on Morse Homology (with an Eye Towards Floer Homology and Pseudoholomorphic Curves)*, available at http://people.math.umass.edu/~sullivan/797SG/hutchings-morse.pdf.

[39] P. Ozsvath, J. Rasmussen and Z. Szabo, *Odd Khovanov homology*, Algebr. Geom. Topol. **13** (2013) 1465–1488; arXiv:0710.4300 [math.QA].

[40] A. P. Ellis and A. D. Lauda, *An odd categorification of $U_q(\mathfrak{sl}_2)$*, Quantum Topol. **7** (2016) 329–433; arXiv:1307.7816 [math.QA].

[41] V. Mikhalylov and E. Witten, *Branes and supergroups*, Commun. Math. Phys. **340** (2015) 699–832; arXiv:1410.1175 [hep-th].

[42] W. Nahm, *A simple formalism for the BPS monopole*, Phys. Lett. **B90** (1980) 413–414.

[43] P. Kronheimer, *Instantons and the geometry of the nilpotent variety*, J. Diff. Geom. **32** (1990) 473–490.

[44] P. Kronheimer, *A hyper-Kahlerian structure on coadjoint orbits of a semisimple complex Lie group*, J. Lond. Math. Soc. **42** (1990) 193–206.

[45] M. F. Atiyah and R. Bielawski, *Nahm's equations, configuration spaces, and flag manifolds*, arXiv:math/0110112 [math.RT].

[46] R. Mazzeo and E. Witten, *The Nahm pole boundary condition*, arXiv:1311.3167 [math.DG].

[47] V. Mikhaylov, *On the solutions of generalized Bogomolny equations*, JHEP **1205** (2012) 112; arXiv:1202.4848 [hep-th].

[48] D. Gaiotto and E. Witten, *Knot invariants from four-dimensional gauge theory*, Adv. Theor. Math. Phys. **16** (2012) 935–1086; arXiv:1106.4789 [hep-th].

[49] D. Gaiotto, G. W. Moore and E. Witten, *Algebra of the infrared: string field theoretic structures in massive $\mathcal{N} = (2,2)$ field theory in two dimensions*, arXiv:1506.04087 [hep-th].

2 Elementary Knot Theory

MARC LACKENBY

2.1 Introduction

In the past 50 years, knot theory has become an extremely well-developed subject. But there remain several notoriously intractable problems concerning knots and links, many of which are surprisingly easy to state. The focus of this chapter is this 'elementary' aspect to knot theory. Our aim is to highlight what we still do not understand, as well as to provide a brief survey of what is known. The problems that we will concentrate on are the non-technical ones, although of course, their eventual solution is likely to be sophisticated. In fact, one of the attractions of knot theory is its wide variety of interactions with many different branches of mathematics: 3-manifold topology, hyperbolic geometry, Teichmüller theory, gauge theory, mathematical physics, geometric group theory, graph theory and many other fields.

This survey and problem list is by no means exhaustive. For example, we largely neglect the theory of braids, because there is an excellent survey by Birman and Brendle [5] on this topic. We have also avoided 4-dimensional questions, such as the slice–ribbon conjecture (Problem 1.33 in [41]). Although these do have a significant influence on elementary knot theory, via unknotting number for example, this field is so extensive that it would best be dealt with in a separate article.

In fact, this chapter comes with a number of health warnings. It is frequently vague, often deliberately so, and many standard terms that we use are not defined here. It is certainly far from complete, and we apologize to anyone whose work has been omitted. And it is not meant to be a historical account of the many developments in the subject. Instead, it is primarily a list of fundamental and elementary problems, together with enough of a survey of the field to put these problems into context.

2.2 The equivalence problem

The equivalence problem for knots and links asks the most fundamental question in the field: *can we decide whether two knots or links are equivalent?* Questions of this sort arise

Lectures on Geometry. Edward Witten, Marc Lackenby, Martin R. Bridson, Helmut Hofer and Rahul Pandharipande.
© Oxford University Press 2017. Published 2017 by Oxford University Press.

naturally in just about any branch of mathematics. In topology, there are many well-known negative results. A central result of Markov [65] states that there is no algorithm to determine whether two closed n-manifolds are homeomorphic when $n \geq 4$. But in dimension 3, the situation is more tractable. In particular, the equivalence problem for knots and links is soluble. In fact, there are now five known ways to solve this problem, which we describe briefly below.[1] Four of these techniques focus on the *exterior* $S^3 - \text{int}(N(K))$ of the link K. Now, one loses a small amount of information when passing to the link exterior, because it is possible for distinct links to have homeomorphic exteriors. (This is not the case for knots, by the famous theorem of Gordon and Luecke [23].) So, one should also keep track of a complete set of meridians for the link—in other words, a collection of simple closed curves on $\partial N(K)$, one on each component of $\partial N(K)$, each of which bounds a meridian disc in $N(K)$. Then, for two links K and K', there is a homeomorphism between their exteriors that preserves these meridians if and only if K and K' are equivalent.

2.2.1 Hierarchies and normal surfaces

Haken [25], Hemion [30] and Matveev [66] were the first to solve the equivalence problem for links. Their solution is lengthy and difficult. A full account of their argument was given only fairly recently by Matveev [66].

Haken used incompressible surfaces arranged into sequences called hierarchies. Recall that a surface S properly embedded in a 3-manifold M is *incompressible* if, for any embedded disc D in M with $D \cap S = \partial D$, there is a disc D' in S with $\partial D' = \partial D$. Given such a surface, it is natural to cut M along S, creating the compact manifold $\text{cl}(M - N(S))$. It turns out that one can always find another properly embedded incompressible surface in this new manifold, and thereby iterate this procedure. A *hierarchy* is a sequence of manifolds and surfaces

$$ M = M_1 \xrightarrow{S_1} M_2 \xrightarrow{S_2} \cdots \xrightarrow{S_{n-1}} M_n $$

where each S_i is a compact orientable incompressible surface properly embedded in M_i with no 2-sphere components, where M_{i+1} is obtained from M_i by cutting along S_i and where the final manifold M_n is a collection of 3-balls. Haken proved that the exterior of a non-split link always admits a hierarchy. He was able to solve the equivalence problem for links via the use of hierarchies with certain nice properties. An example of a hierarchy for a knot exterior is given in Fig. 2.1.

Associated with a hierarchy, there is a 2-complex embedded in the manifold M. To build this, one extends each surface S_i in turn, by attaching a collar to ∂S_i in $N(S_1 \cup \cdots \cup S_{i-1})$, so that the boundary of the new surface runs over the earlier surfaces and ∂M. The union of these extended surfaces and ∂M is the 2-complex. The 0-cells are the points where three surfaces intersect. The 1-skeleton is the set of points that lie on at least two surfaces. Since the complement of this extended hierarchy and the boundary of the manifold is a collection of open balls, we actually obtain a cell structure for M in this way. It is a trivial but important observation that this is a representation of M that depends only on

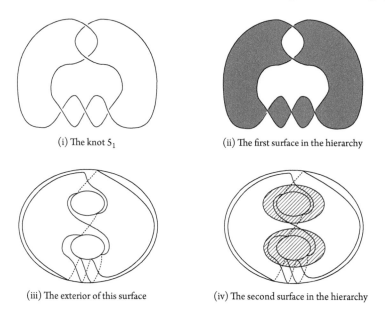

(i) The knot 5_1

(ii) The first surface in the hierarchy

(iii) The exterior of this surface

(iv) The second surface in the hierarchy

Figure 2.1 A hierarchy for a knot exterior.

the topology of the hierarchy. Haken's solution to the equivalence problem establishes that, as long as one chooses the hierarchy correctly, there is a way of building this cell structure, starting only from an arbitrary triangulation of M.

Suppose that we are given diagrams for two (non-split) links K and K'. The first step in the algorithm is to construct triangulations T and T' for their exteriors M and M'. We know that M and M' admit hierarchies with the required properties. Moreover, if M and M' are homeomorphic (by a homeomorphism preserving the given meridians), then this homeomorphism takes one such hierarchy for M to a hierarchy for M'. So, let us focus initially on the hierarchy for M. The next stage in the procedure is to place the first surface S of the hierarchy into *normal form*. By definition, this means that it intersects each tetrahedron of T in a collection of triangles and squares, as shown in Fig. 2.2.

Now, any incompressible, boundary-incompressible surface S (with no components that are 2-spheres or boundary-parallel discs) in a compact orientable irreducible boundary-irreducible 3-manifold M with a given triangulation T may be ambient isotoped into normal form. It can then be encoded by a finite amount of data. The normal surface S determines a collection of $7t$ integers, where t is the number of tetrahedra in T. These integers simply record the number of triangles and squares of S in each tetrahedron. (Note from Fig. 2.2 that there are 4 types of normal triangle and 3 types of normal square in each tetrahedron.) The list of these integers is denoted by $[S]$ and is called the associated *normal surface vector*. Normal surface vectors satisfy a number of constraints. First, each coordinate is, of course, a non-negative integer. Second, they satisfy various linear equations, called the *matching conditions*, that guarantee that, along each face of T with a tetrahedron on each side, the triangles and squares of S in the two adjacent

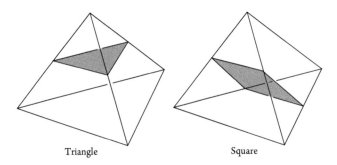

Triangle   Square

Figure 2.2 Normal surface.

tetrahedra patch together correctly. Finally, there are the *quadrilateral constraints*, which require S to have at most one type of square in each tetrahedron. This is because two different square types cannot coexist in the same tetrahedron without intersecting. Haken showed that there is a one-to-one correspondence between properly embedded normal surfaces (up to a suitable notion of 'normal' isotopy) and vectors satisfying these conditions. Given this strong connection, it is natural to use tools from linear algebra when studying normal surfaces. In particular, one can speak of a normal surface S as being the *sum* of two normal surfaces S_1 and S_2 if $[S] = [S_1] + [S_2]$. A normal surface S is called *fundamental* if it cannot be expressed as a sum of two other non-empty normal surfaces. Haken showed that there is a finite list of fundamental normal surfaces, and these are all algorithmically constructible. A key part of Haken's argument is to show that some surfaces can be realized as fundamental normal surfaces with respect to *any* triangulation of the manifold. Thus, if we could find a hierarchy where each surface had this property, then we could construct the hierarchy starting with any triangulation T of M. In fact, Haken did not prove that there is a hierarchy with this property, but he did show that there is a hierarchy where each surface can be realized as a sum of boundedly many fundamental normal surfaces. This is sufficient to make the argument work.

Thus, Haken's algorithm essentially proceeds by starting with the triangulations T and T' of the two link exteriors. Then, one constructs all possible hierarchies for each manifold, with the property that each surface in the hierarchy is a bounded sum of fundamental surfaces. From each such hierarchy, one forms the associated cell structure for M. If two such cell structures, one from each of T and T', are combinatorially equivalent, then the links are the same. If none of these cell structures are equivalent, then the links are distinct.

There were two cases that Haken could not handle using these methods. When M fibres over the circle with fibre S, then cutting M along S results in a copy of $S \times I$. In $S \times I$, there is no good choice of surface to cut along. In particular, it is not clear how to find a surface that is guaranteed to be fundamental or at least a bounded sum of fundamental surfaces. The resolution of the equivalence problem for these manifolds was not completed until work of Hemion [30], where he gave a solution to the conjugacy problem in the mapping class group of S. A similar problem arises when cutting along S

yields a twisted I-bundle (in which case, S is known as a *semi-fibre*). This case was fully dealt with by Matveev [66].

In fact, there is a now an alternative way of sidestepping this issue, by establishing that there is always a hierarchy for a non-split link exterior that is always constructible. This relies on a result of Culler and Shalen [13], which says that if M is a compact orientable hyperbolic 3-manifold with boundary a non-empty collection of tori, then M contains either a closed, non-separating, essential properly embedded surface or a separating, connected, essential, properly embedded surface with non-empty boundary. When M is the exterior of a hyperbolic link in the 3-sphere, this surface can be chosen to be neither a fibre nor a semi-fibre (see Section 7 in [9]). By work of Mijatovic [71], there is such a surface that can be realized as a bounded sum of fundamental surfaces.

Of course, this summary is a dramatic over-simplification. The details of the argument are very complicated. A full exposition, which takes several hundred pages, can be found in the excellent book by Matveev [66].

2.2.2 Geometric structures

One of the most striking advances in low-dimensional topology was Thurston's introduction of hyperbolic geometry, via his Geometrization Conjecture [100]. This was proved by Perelman in 2003 [81–83], but Thurston himself proved his conjecture in the case of knot and link complements, and this can be used to provide another method of solving the equivalence problem. An approximate statement of the Geometrization Conjecture is that every compact orientable 3-manifold admits a 'canonical decomposition' into 'geometric pieces'.

The canonical decomposition takes place in two steps. (We focus for simplicity on the case where the boundary of the 3-manifold is empty or a collection of tori.) First, one decomposes the 3-manifold into its prime connected summands. Second, one cuts the manifold along its JSJ tori. Both processes can be achieved algorithmically using normal surface theory [34]. (See [75] and [66] for more details of JSJ theory, which was first developed by Jaco, Shalen [33] and Johannson [36].)

Once this decomposition has taken place, the resulting pieces either are Seifert-fibred or admit finite-volume hyperbolic structures. The generic situation is the hyperbolic case, and it is this that we will focus on. An algorithm to find a finite-volume hyperbolic structure on a 3-manifold, provided one exists, was given by Manning [63], based on work of Casson. By combining [63] with work of Weeks [103], it possible to compute the Epstein–Penner decomposition of the manifold [18]. This is a striking construction: it is a decomposition of a non-compact finite-volume hyperbolic 3-manifold into ideal polyhedra. The crucial thing about it is that it is canonical: it depends only on the topology of the manifold, not on any arbitrary choices. Thus, two finite-volume non-compact hyperbolic 3-manifolds are homeomorphic if and only if their Epstein–Penner decompositions are combinatorially equivalent. Moreover, one can determine whether there is a homeomorphism between the manifolds, taking one given collection of disjoint simple closed curves in the boundary to another. This therefore gives a solution to the equivalence problem for knots and links.

This is more than just a theoretical algorithm. The computer programs Knotscape [31], Snappea [102] and Snap [10] all attempt to compute the Epstein–Penner decomposition of a hyperbolic 3-manifold. Unlike the Casson–Manning algorithm, they are not guaranteed to do so. But *if* Snap or Knotscape finds what it claims is the Epstein–Penner decomposition, then this is indeed the correct decomposition, and, as a result, these programs are a practical method of reliably determining whether two links are equivalent.

2.2.3 Relatively hyperbolic groups

One can also use some of the machinery of geometric group theory to solve the equivalence problem, at least for hyperbolic knots.

The fundamental group of a closed hyperbolic 3-manifold is Gromov-hyperbolic. Mostow rigidity states that two such manifolds are isometric if and only if their fundamental groups are isomorphic. Moreover, Sela provided an algorithm to decide whether two torsion-free Gromov-hyperbolic groups are isomorphic [92]. Now, the complement of a knot is, of course, not closed. But when this complement has a hyperbolic structure, its fundamental group is a relatively hyperbolic group, relative to the fundamental group of the toral boundary. Mostow rigidity applies in this case also. Sela's result was extended to this class of relatively hyperbolic groups by Dahmani and Groves [14]. So, at least in the case of hyperbolic knots, one can solve the equivalence problem this way. This does not immediately lead to a solution for hyperbolic links, because links are not determined by their complements. But presumably, the algorithm of Dahmani and Groves could be adapted to take account of a set of meridians in the boundary tori. This method also only works in the hyperbolic case, and so to turn this into a fully fledged solution to the equivalence problem for all links, presumably one would first need to find the 'canonical decomposition into geometric pieces' that was discussed in Section 2.2.2. Nevertheless, this is a genuinely different solution to the equivalence problem. But its algorithmic complexity seems to be hard to estimate.

2.2.4 Reidemeister moves

There is an alternative way of interpreting the equivalence problem in terms of Reidemeister moves. Recall that a *Reidemeister move* is a local modification to a link diagram as shown in Fig. 2.3. Reidemeister proved [85] that any two diagrams of a link differ by a sequence of Reidemeister moves. The algorithmic significance of this is that *if* two diagrams represent the same link, then one may always prove this, given enough time and computing power. This is because one can apply all possible Reidemeister moves to the first diagram, and thereby obtain a collection of diagrams of the first link. One can then

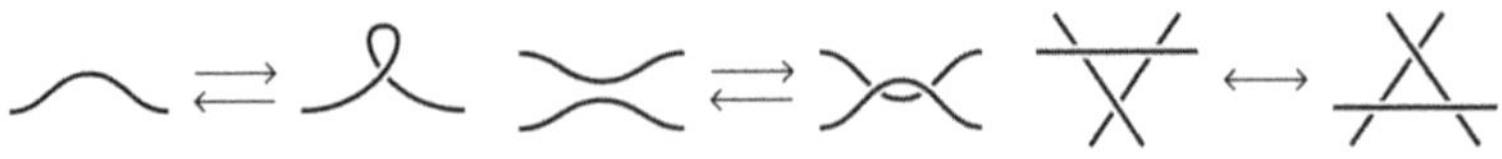

Figure 2.3 Reidemeister moves.

apply all possible Reidemeister moves to each of these, and so on. It is a consequence of Reidemeister's theorem that if the two initial diagrams represent the same link, then this process is guaranteed to produce a sequence of Reidemeister moves taking one diagram to the other.

Of course, this does not provide a solution to the equivalence problem, because if two diagrams represent distinct link types, then the above process does not terminate. But if one knew in advance how many Reidemeister moves were required to take one diagram to other, then one could stop the process when one had tried all possible sequences of Reidemeister moves of this length, and if a sequence of Reidemeister moves taking one diagram to the other had not been found, then one could declare that the links are distinct. More specifically, a computable upper bound on the number of Reidemeister moves required to relate two diagrams of the same link provides a solution to the equivalence problem. In fact, it is not hard to show that the converse is also true: if there is a solution to the equivalence problem, then, given natural numbers n_1 and n_2, one can compute an upper bound on the number of Reidemeister moves required to relate two diagrams of a link with n_1 and n_2 crossings. One enumerates all link diagrams with these numbers of crossings and then sorts them into their various link types, using the hypothesized algorithm to solve the equivalence problem. Then, for all diagrams of the same link type, one starts applying Reidemeister moves to these diagrams. By Reidemeister's theorem, eventually a sequence of such moves will be found relating any two diagrams of the same link. Hence, one can compute an upper bound on the number of moves that are required.

Recently, Coward and the author [9] have provided an explicit, computable upper bound on Reidemeister moves, thereby giving a new and conceptually simple solution to the equivalence problem. Unfortunately, it is a huge bound, and so the resulting algorithm is very inefficient. We will provide more details in Section 2.4, which is devoted to Reidemeister moves.

2.2.5 Pachner moves

It is well known that two triangulated manifolds are PL-homeomorphic if and only if they differ by a sequence of Pachner moves [80]. These are modifications that change the triangulation in a very simple way. In the case of closed 3-manifolds, the moves are shown in Fig. 2.4. For 3-manifolds with boundary, there are some extra moves that change the triangulation on the boundary.

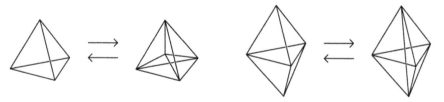

Figure 2.4 Pachner moves.

Therefore, just as argued in Section 2.2.4, if two triangulated manifolds are PL-homeomorphic, then one will always be able to prove this eventually. Moreover, if one has a computable upper bound on the number of Pachner moves required to pass between two triangulations, then one can solve the PL-homeomorphism problem. In the case of knot exteriors, such a computable upper bound was found by Mijatovic [71]. Unfortunately, his bound is massive: a tower of exponentials, exponentially high. So, again, this does not lead to a practical algorithm.

Mijatovic's method built on work of Haken. As explained in Section 2.2.1, Haken used a hierarchy for a link exterior M to build a 'canonical' cell structure for M. This had the property that one can algorithmically build this cellulation starting from any triangulation of M. By carefully modifying the initial triangulation to one derived from this cellulation, Mijatovic was able to produce an upper bound on the number of Pachner moves required. The method is not easy, and in particular, he needed to go beyond what Haken achieved, because he also needed the Rubinstein–Thompson machinery for 3-sphere recognition [70, 99].

2.2.6 Solving the equivalence problem efficiently

None of these approaches to the equivalence problem is known to be efficient. Indeed, any efficient solution seems to be out of reach at present. This leads to our first unsolved problem.

Problem 2.1. *What is the complexity class of the equivalence problem for knots and links?*

It is quite striking that, of the five above approaches to the equivalence problem, only the final two give an a priori estimate for its running time. However, the disadvantage of the approaches using Reidemeister moves or Pachner moves is that they are almost inevitably lengthy.

One the main difficulties in Haken's approach to the equivalence problem is that, in order to apply normal surface theory, one needs to build a triangulation T_i for each 3-manifold M_i in the hierarchy. However, the surface S_i may be exponentially complicated in T_i, in the sense of having exponentially many triangles and squares, as a function of the number of tetrahedra in T_i. So, it seems inevitable that the next triangulation T_{i+1} should be exponentially more complicated than that of T_i. So, the number of tetrahedra in T_n is massive. As a function of the crossing number of a given diagram of the link, it is a tower of exponentials. The height of this tower is n, the length of the hierarchy, and unfortunately, owing to the technical requirements of Haken's hierarchies, n can be quite large too.

On the other hand, the approach to the equivalence problem using geometric structures is of unknown complexity. However, it appears to be computationally efficient in practice, even if one cannot yet prove this. So, there remains some hope that a provably efficient solution to the equivalence problem may be found in the future.

One might ask for a polynomial-time algorithm in Problem 2.1. But this seems rather unlikely, and there certainly seems no prospect of a proof in sight. Even an NP algorithm seems beyond reach at the moment. (For a very brief explanation of NP and other complexity classes, see Section 2.3.1.) Alternatively, one might try to find lower bounds on the complexity of the problem, conditional upon well-known conjectures in theoretical computer science. For example, is the equivalence problem NP-hard? In other words, is it at least as hard as any other NP problem? In particular, if there were a polynomial-time solution to the equivalence problem for knots and links, would this imply that P = NP? This seems plausible. Indeed, a seemingly simpler algorithm in 3-manifold theory is known to be NP-complete, by work of Agol, Hass and Thurston [3]. This is the problem of determining whether a given knot in a given 3-manifold bounds a compact orientable embedded surface with genus at most a given integer.

2.3 The recognition problem

An algorithmic problem that is closely related to the equivalence problem is the *recognition problem* for a fixed link type. Here, one fixes a link type K, and one asks whether a given diagram represents this link type. Fairly obviously, a solution to the equivalence problem implies a solution to the recognition problem for each link type. However, by fixing the link type, we simplify the problem, and so there may be some hope of coming up with a more efficient solution. The case of the unknot is particularly intriguing. The following remains unknown.

Problem 2.2. *Is there a polynomial-time algorithm to recognize the unknot?*

Currently, the unknot recognition problem is known to lie in the following complexity classes: NP, co-NP and E. We briefly remind the reader of the definitions of these and various other classes.

2.3.1 Some basic complexity classes

The class P is possibly the simplest to describe. It consists of those problems that can be solved in time n^k, where n is the size of the input and k is a fixed constant. Similarly, a problem is in the class E if it can be solved in time k^n, where again n is the size of the input and k is a fixed constant. Somewhat confusingly, this is not the same as the class EXP of exponential-time algorithms, since, by definition, these can solved in time k^{n^c} for constants k and c.

A problem is in NP if it admits a polynomial-time certificate. This is an extra piece of information that is not provided by the algorithm, but rather is given by some external source. The point is that *if* the answer to the problem is 'yes', then the certificate proves that it is 'yes', and this can be verified in polynomial time. There are many examples of NP problems; one is the problem of deciding whether a given positive integer is composite. In this case, a very simple certificate is two integers (greater than 1) whose product is the given integer. It can be verified in polynomial time (as a function of the number of digits of the input) that these two integers do indeed multiply together to produce the given number. One might legitimately wonder why NP problems are solvable at all, given

that they require information from an external source. But the point is that this information must have polynomially bounded size, because a computer has time enough only to
check this many bits of information. So, a deterministic algorithm proceeds by checking
all possible certificates of at most this size. Thus, problems in NP are also in EXP.

A computational problem is in co-NP if its negation is in NP. Thus, for example, the
problem of deciding whether a positive integer is composite is in co-NP, because there
is an NP algorithm to determine whether a positive integer is prime. This is not at all
obvious however. It was first proved by Pratt [84], using the Lucas primality test.

There is also the well-known class of NP-complete problems. A problem is
NP-complete if it is in NP and any other NP problem may be (efficiently) reduced to
it. In particular, if an NP-complete problem has a polynomial-time solution, then this
would imply that P = NP. It is widely believed that P $\neq$ NP, and so NP-complete problems are, in some conditional sense, provably hard. However, it is very unlikely that any
problem that is both in NP and co-NP is NP-complete, because this would imply that
NP = co-NP, which also is viewed as rather unlikely.

2.3.2 Haken's unknot recognition algorithm

The first person to provide an unknot recognition algorithm was Haken [24]. A knot K
is the unknot if and only if its exterior M has compressible boundary. Haken showed
how one may use normal surface theory to determine this. As in Section 2.2.1, one
builds a triangulation of M using the given diagram of K. Haken showed that if M contains an essential properly embedded disc, then it contains one that is a fundamental
normal surface. He also showed that all fundamental normal surfaces may be algorithmically constructed. Thus, his unknot recognition algorithm proceeds by constructing
all possible fundamental normal surfaces, and determining whether any of these is an
essential disc.

2.3.3 The NP algorithm of Hass, Lagarias and Pippenger

Hass, Lagarias and Pippenger [27] provided an NP algorithm by building on Haken's
work. Instead of using fundamental surfaces, they used vertex surfaces. By definition, a
normal surface S is a *vertex surface* if it is connected and, whenever $n[S] = [S_1] + [S_2]$ for
a positive integer n and normal surfaces S_1 and S_2, then each of $[S_1]$ and $[S_2]$ is a multiple
of $[S]$. The reason for the vertex terminology is that the set of all normal surface vectors
can be viewed as integer points within a larger subset of $\mathbb{R}^{7t}$, where t is the number of
tetrahedra of the 3-manifold M. This subset is the set of vectors satisfying the matching equations and quadrilateral constraints, but where the vectors are required only to
be non-negative *real* numbers. It is not hard to see that this subset is a union of convex
polytopes, each of which is the cone on a compact polytope, where the cone point is the
origin. Vertex surfaces are precisely those connected normal surfaces that are a multiple
of a vertex of one of these compact polytopes. Hass, Lagarias and Pippenger showed that
if M contains an essential properly embedded disc, then it has one that is a vertex normal
surface. Moreover, for each such vertex surface that is an essential disc, one can certify
that it is an essential disc in polynomial time. Thus, unknot recognition lies in NP. It is

also not hard to prove that the number of vertex surfaces is at most 2^{7t}, because each one is obtained as the unique solution (up to scaling) of the matching equations plus some extra constraints that force certain coordinates to be zero. So, if one checks each such vertex surface in turn, one can determine whether the given knot is the unknot in at most k^n steps, where n is the crossing number of the diagram and k is a fixed constant. This implies that unknot recognition lies in E.

2.3.4 The co-NP algorithm of Kuperberg

Kuperberg has recently proved that unknot recognition is in co-NP, assuming the Generalized Riemann Hypothesis [48]. Here, one wants an efficient way of certifying that a non-trivial knot K is indeed knotted. Kuperberg achieves this by establishing the existence of a representation from $\pi_1(S^3 - K)$ to some finite group with non-abelian image. The unknot clearly admits no such representation, because the fundamental group of its complement is $\mathbb{Z}$. Thus, this does establish that the knot is non-trivial.

The existence of such a representation is not at all clear. Moreover, Kuperberg requires a representation with a small enough image that the representation can be verified as a homomorphism with non-abelian image in polynomial time. The starting point is the major theorem of Kronheimer and Mrowka [45] that establishes that, for a non-trivial knot K, there is a homomorphism $\pi_1(S^3 - K) \to SU(2)$ with non-abelian image. This is established using a highly sophisticated argument that uses taut foliations, contact structures, symplectic fillings and the instanton equations. (In the case where the knot K is hyperbolic, one can instead use the associated representation $\pi_1(S^3 - K) \to PSL(2, \mathbb{C})$ in Kuperberg's argument.)

Once one has a linear representation of $\Gamma = \pi_1(S^3 - K)$, there is a standard method of obtaining finite representations that goes back to the result of Malce'ev [62] that states that finitely generated linear groups are residually finite. One considers a finite generating set for Γ and, for each generator, one considers the entries of the corresponding matrix (in $SU(2)$). The ring R generated by these entries is a finitely generated subring of $\mathbb{C}$. Thus, the image of Γ lies in $SL(2, R)$. Such a ring R contains many finite-index ideals I. The desired finite quotients are obtained by mapping $SL(2, R)$ onto $SL(2, R/I)$ for some such I. The Generalized Riemann Hypothesis is used to prove the existence of an ideal I with small index, and hence one obtains a finite quotient of Γ with small size. The actual tools that are used here are results of Korain [43], Lagarias-Odlyzko [59] and Weinberger [104] that imply that, for any integer polynomial, one may reduce its coefficients modulo some small prime and obtain a polynomial with a root, assuming the Generalized Riemann Hypothesis.

2.3.5 The co-NP algorithm of Agol

Agol has announced the existence of another certificate for establishing that a non-trivial knot is knotted [1]. This has the advantage that it is unconditional on any conjectures, and it also provides more information about the knot, by determining its genus. Recall that the *genus* of a knot K is the minimal genus of a Seifert surface for K. It is a fundamental quantity, but, for the purposes of this algorithm, all that one needs is that the genus is zero if and only if the knot is trivial.

It is a consequence of Haken's work that the genus of a knot is algorithmically comput-able, because a minimal-genus Seifert surface can be arranged to be a fundamental normal surface. However, Thurston and Gabai found another method for determining the genus of knots, by using the theory of taut foliations [21, 101]. In fact, this method natur-ally measures not the genus of a compact surface S but its *Thurston complexity* $\chi_-(S)$. By definition, when S is connected, then $\chi_-(S) = \max\{-\chi(S), 0\}$. When S is disconnec-ted with components $S_1, \ldots, S_n$, then $\chi_-(S) = \sum_{i=1}^n \chi_-(S_i)$. Thus, when S is orientable, $\chi_-(S)$ is roughly the negative of the Euler characteristic of the surface, but one first dis-cards disc and sphere components. So, for knots, a Seifert surface has minimal possible χ_- if and only if it has minimal genus. But this need not be the case for links, because the link may have disconnected Seifert surfaces as well as connected ones.

The theory of Thurston and Gabai is concerned with foliations on a 3-manifold, where the leaves of the foliation have codimension 1 and where there is a consistent transverse orientation on these leaves. Such a foliation is *taut* if there is a simple closed curve in the 3-manifold that is transverse to the foliation and that intersects every leaf. This may seem a slightly strange definition, but it turns out that the existence of such a curve rules out various trivial examples of foliations, and in fact the following was proved by Thurston [101]: *If S is a compact leaf of a taut foliation on a compact orientable 3-manifold M, then S has minimal χ_- in its homology class in $H_2(M, \partial S)$.* Hence, if one starts with a Seifert surface S for an oriented link L, and one can find a taut foliation on the exterior of L in which S is a compact leaf, then S has minimal χ_- among all Seifert surfaces for L. Gabai [21] introduced a method for constructing these foliations and in fact proved the con-verse statement: *If S is a Seifert surface for an oriented non-split link L with minimal χ_-, then there is some taut foliation on the exterior of L in which S is a leaf.*

One can view this taut foliation as a way of certifying the genus of a knot. However, it is not a certificate in the strict algorithmic sense, because, in principle, an infinite amount of information is required to specify the foliation. But the key features of Gabai's foli-ations are encoded by a finite amount of information as follows. Gabai utilized hierarchies to construct his foliations, and it was first observed by Scharlemann [88] that many of the useful topological consequences of Gabai's theory can be extracted solely from the hierarchy, without referring to the foliation. The hierarchies of Gabai and Scharlemann naturally occur in the setting of 3-manifolds with the following extra structure.

A *sutured manifold* is a compact oriented 3-manifold M with its boundary decomposed into two subsurfaces R_- and R_+. A transverse orientation is imposed on these subsur-faces, so that R_+ points out of M and R_- points into it. These surfaces intersect along a collection of simple closed curves γ, called *sutures*. It is usual to denote a sutured manifold by (M, γ), but the surfaces R_+ and R_- are also an essential part of the struc-ture. A sutured manifold (M, γ) is *taut* if M is irreducible, R_+ and R_- are incompressible and they minimize χ_- in their homology classes in $H_2(M, \gamma)$.

When a sutured manifold (M, γ) is decomposed along a properly embedded, trans-versely oriented surface S that intersects γ transversely, the resulting 3-manifold M' inherits a sutured manifold structure. This decomposition is *taut* if M and M' are both taut as sutured manifolds and S has minimal χ_- in its homology class in $H_2(M, \partial S)$. A sequence of such decompositions terminating in a collection of 3-balls is a *sutured manifold hierarchy*. A key feature of sutured manifold theory is the following: *If a sequence of sutured manifold decompositions ends in a collection of taut sutured 3-balls, then (provided some simple conditions hold) every decomposition in this sequence is taut.*

This sequence of decompositions can therefore be viewed as a certificate for knot genus. However, it is not at all clear that one can produce such a certificate that can be verified in polynomial time. The principal difficulty is the one explained in Section 2.2.6 in reference to Haken's use of hierarchies. It seems, at first sight, that one needs to keep track of a triangulation of each manifold in this hierarchy, and the complexity of these triangulations seems to grow rapidly as one proceeds along the hierarchy. Agol's alternative approach is to make the entire sutured manifold hierarchy normal (in some suitable sense) with respect to the initial triangulation of M. The idea is that the sutured manifold hierarchy is closely related to a taut foliation and so normalizing such a hierarchy is somewhat analogous to normalizing a taut foliation. Techniques exist for placing taut foliations in 'normal' form, due to Brittenham [8] and Gabai [22], and versions of these are used by Agol. Unfortunately, full details of this proof have not yet been written down. An alternative approach may be to use the techniques of the author in [50], where sutured manifold hierarchies were normalized in some weak sense.[2]

2.3.6 Khovanov homology

In the mid 1980s, knot theory underwent a dramatic revolution, with the introduction of the Jones polynomial [37], and then Witten's interpretation of this using Chern–Simons theory [105]. Right from the outset of this work, it was asked: *Does the Jones polynomial detect the unknot?* In other words, must a non-trivial knot necessarily have non-trivial Jones polynomial? This question remains unanswered. Indeed, it may be viewed as a central open problem in knot theory, but it is not 'elementary' and so does not make it onto our problem list.

It was shown recently by Kronheimer and Mrowka [47] that a more refined invariant, Khovanov homology, does detect the unknot. This followed from a slightly indirect relationship between Khovanov homology and 4-dimensional gauge-theoretic invariants.

It seems unlikely that this result will lead to an efficient way of recognizing the unknot. This is because determining Khovanov homology is a computationally intensive process. In fact, the less refined Jones polynomial is, in one sense, provably hard to calculate. For example, it was shown by Jaeger, Vertigana and Welsh [35] that computation of the Jones polynomial of a link is '$\sharp$P-hard'. (This is somewhat similar to being NP-hard.)

2.3.7 Heegaard Floer homology

Another major advance in knot theory has been the introduction of Heegaard Floer homology by Oszváth and Szabó [77]. This was based on earlier instanton invariants, as pioneered by Donaldson [16] in dimension 4 and then by Floer [19] in dimension 3, and monopole invariants, as developed by Seiberg and Witten [106] in dimension 4 and then by Kronheimer and Mrowka [46] in dimension 3. It was clear from the outset that these gauge-theoretic invariants were particularly powerful, but they were undoubtedly hard to calculate. One of the most attractive features of Heegaard Floer homology is that, right from its inception, it has been possible to calculate it in many important examples. Indeed, there are now algorithms to calculate the various different versions of Heegaard Floer homology, due to Sarkar and Wang [86] and Manolescu, Ozsváth and Sarkar [64]. One of the earliest results about Heegaard Floer homology was that it detects the unknot [78]. So, coupled with the fact that it is computable, this provides another unknot recognition algorithm. Again, however, it seems unlikely that this algorithm is in any way efficient.

2.3.8 Rectangular diagrams and arc presentations

In 2003, Dynnikov introduced a new and striking solution to the unknot recognition problem [17]. This used rectangular diagrams and arc presentations. A *rectangular diagram* (also called *grid diagram*) is a planar link diagram that consists of vertical and horizontal arcs, no two of which are colinear, and with the key property that whenever a horizontal and a vertical arc cross, it is the vertical arc that is the over-arc at the resulting crossing. The number of vertical arcs is equal to the number of horizontal arcs; this is known as the *arc index* of the rectangular diagram (Fig. 2.5).

Rectangular diagrams were studied in detail by Cromwell [12]. He introduced a set of moves that change a rectangular diagram, without changing the link type, much like Reidemeister moves. He showed that any two rectangular diagrams for a link differ by a sequence of such moves. (See Fig. 2.6.) Two of these types of moves (*cyclic permutations* and *exchange moves*) leave the arc index unchanged. The other type of move is a *stabilization/destabilization*, which increases/decreases the arc index by one.

Dynnikov proved that any rectangular diagram of the unknot can be reduced to the trivial diagram using these moves, but, crucially, *no stabilizations are required*. This leads to the following unknot recognition algorithm. Start with the given diagram, and, using

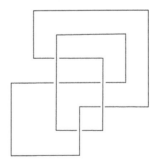

Figure 2.5 A rectangular diagram.

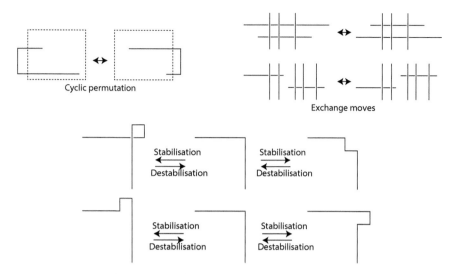

Figure 2.6 Modifications to a rectangular diagram.

an isotopy of the plane, make it rectangular. Then keep applying all possible cyclic permutations, exchange moves and destabilizations. Only finitely many moves are possible at each stage. The key point is that the number of rectangular diagrams with bounded arc index is finite. So, eventually, one can explore the entire set of rectangular diagrams that are reachable from the initial one by sequences of these moves. The given knot is the unknot if and only if the trivial diagram is in this collection.

This does not immediately lead to an efficient solution. This is because the number of rectangular diagrams with a given arc index is a super-exponential function, and so the search space might be quite large. But it is a beautiful and striking result that is perhaps the closest in spirit to elementary knot theory among all the known unknot recognition algorithms. It also has the following consequence, as noted by Dynnikov himself: *If a diagram of the unknot has n crossings, then there is a sequence of Reidemeister moves taking it to the trivial diagram, where each diagram in this sequence has crossing number at most* $2(n + 1)^2$. This is simply because a rectangular diagram with arc index m, say, has crossing number at most $(m - 1)^2/2$.

Dynnikov's proof builds on previous work of Cromwell [12], Birman, Menasco [7] and Bennequin [4]. Cromwell observed a correspondence between rectangular diagrams and arc presentations, which are defined as follows. One fixes a round unknot in the 3-sphere, which is a reference circle, known as the *binding circle*. The complement of the binding circle is an open solid torus, which is foliated by open discs called *pages*. A link is in an *arc presentation* if it intersects the binding circle in finitely many points called *vertices*, and it intersects each page in either the empty set or a single open arc joining distinct vertices. Cromwell proved that there is a one-to-one correspondence between arc presentations and rectangular diagrams up to cyclic permutation.

Thus, Dynnikov's proof mainly concentrates on arc presentations. Given an arc presentation of the unknot, its spanning disc can be arranged so that its intersection with the pages induces a singular foliation on the spanning disc. The singularities occur when the disc slices through the binding circle, and also at Morse-type singularities in the complement of the binding circle. Dynnikov's measure of complexity for the disc is the number of singularities of this foliation. Using an argument that relies on the fact that the disc has positive Euler characteristic, Dynnikov showed that this singular foliation must have, at some point, a certain configuration. He then showed that this implies the existence of some exchange moves, cyclic permutations and possibly a destabilization, after which the arc index has been reduced or the number of singularities of the disc has been reduced. Thus, this process is guaranteed to terminate. In particular, it is possible to simplify the disc without using stabilizations. Once the disc is as simple as possible, so too is the rectangular diagram.

This style of argument was very much based on earlier work of Cromwell, Birman, Menasco and Bennequin. For example, Birman and Menasco studied representations of the unknot as a closed braid, and proved that it may be reduced to the trivial braid (of braid index 1) by a sequence of moves, also known as exchange moves and destabilizations. Thus, the braid index does not increase during this sequence of moves. However, this does not immediately lead to an unknot recognition algorithm, because, in the case of braids, there are infinitely many possible exchange moves that can be applied at each stage, and the number of braids with a given braid index is infinite. Nevertheless, Birman and Hirsch [6] did use braids to produce an unknot recognition algorithm, by combining these singular foliation techniques with Hass and Lagarias's work on normal surfaces [26].

2.4 Reidemeister moves

Reidemeister proved that any two diagrams of a link differ by a sequence of Reidemeister moves [85]. This result has many applications. For example, it is often used to show that an invariant of link diagrams actually leads to an invariant of knots and links, by showing that the invariant is unchanged under each Reidemeister move. It also leads to many natural and interesting questions, including the following.

Problem 2.3. *Find good upper and lower bounds on the number of Reidemeister moves required to relate two diagrams of a knot or link.*

In addition to being natural and attractive, this question has algorithmic implications. As explained in Section 2.2.4, there exists a computable upper bound on Reidemeister moves if and only if there exists a solution to the equivalence problem. Moreover, if one just considers diagrams of a fixed link type, then the existence of a computable upper bound on Reidemeister moves is equivalent to the existence of a solution to the recognition problem for that link type. Obviously, one would like upper bounds that are as small as possible, because this leads to more efficient algorithms.

We now give a summary of what is known here. The gap between the known upper and lower bounds are quite startling.

2.4.1 Lower bounds

There is, of course, a linear lower bound on the number of Reidemeister moves required to relate two diagrams of a link. More precisely, if two diagrams have n_1 and n_2 crossings, then the number of moves relating them is at least $|n_1 - n_2|/2$, simply because each Reidemeister move changes the crossing number by at most 2.

Hass and Nowik [28] proved that, in general, at least quadratically many moves may be needed. For each link type, they produced a sequence of diagrams D_n where the crossing number of D_n grows linearly with n but where the minimal number of Reidemeister moves relating D_n to D_0 is a quadratic function of n. They did this by finding an invariant of knot diagrams that does not lead to a knot invariant, but changes in a controlled way when a Reidemeister move is performed. Hence, a 'large' difference in the invariants of two diagrams implies that they necessarily differ by a long sequence of Reidemeister moves.

2.4.2 Upper bounds for general knots and links

In contrast to these lower bounds, the known upper bounds on Reidemeister moves are vast. The best known bound in general is due to Coward and the author [9]: *If two diagrams of a link have n and n' crossings respectively, then these diagrams differ by a sequence of at most*

$$\left. 2^{2^{2^{\cdot^{\cdot^{2^{(n+n')}}}}}} \right\} \text{ height } c^{(n+n')}$$

Reidemeister moves, where c = 10^{10^6}. This is huge, but it was the first known upper bound that is primitive recursive.

The proof relied on Mijatovic's upper bound on Pachner moves between triangulations of a link exterior, as described in Section 2.2.5, which in turn built on Haken's work. Mijatovic's upper bound was a tower of exponentials, much like the formula above. However, it is not at all clear how one can go from a bound on Pachner moves to a bound on Reidemeister moves. The idea is to use the sequence of Pachner moves to build an explicit homeomorphism from the 3-sphere to itself, taking the link arising from the first diagram to the link arising from the second diagram. This homeomorphism is isotopic to the identity (via the 'Alexander trick'), and so one obtains a one-parameter family of links interpolating between the two given ones. Projecting these links gives a one-parameter family of diagrams. The control that one has over the homeomorphism of the 3-sphere can be used to control the complexity of these link projections, and so one obtains an upper bound on the crossing number of all the intermediate diagrams. This in turn can be used to bound the number of Reidemeister moves (as explained in more detail in Section 2.4.5).

2.4.3 Upper bounds for the unknot

A recent theorem of the author [55] provides a polynomial upper bound on Reidemeister moves for the unknot: *If a diagram of the unknot has n crossings, then there is a sequence of at most $(236n)^{11}$ Reidemeister moves taking it to the trivial diagram.*

Previously, the best known upper bound was an exponential function of n. This was due to Hass and Lagarias [26]. Their method of proof relied heavily on Haken's solution of the recognition problem for the unknot. One is given a diagram of the unknot with n crossings, and from this one builds a triangulation of the unknot's exterior. It is not hard to arrange that each tetrahedron of this triangulation is straight in $\mathbb{R}^3$ (apart from the tetrahedron enclosing the point at infinity) and that the number of tetrahedra is bounded above by a linear function of n. Now a spanning disc for the unknot can be placed in normal form, and can in fact be realized as a vertex surface. Crucially, there is an upper bound on the number of triangles and squares in a vertex normal surface, which is an exponential function of the number of tetrahedra in the triangulation. One then simply slides the unknot along this disc. Each time one slides across a triangle or square, this induces a controlled number of Reidemeister moves in the projection to the plane of the diagram.

So, it is the exponential upper bound on the size of the normal spanning disc that leads to the Hass–Lagarias upper bound on the number of Reidemeister moves. It is possible to prove that the exponential upper bound on the size of the spanning disc cannot be improved upon. Hass, Snoeyink and Thurston [29] found a sequence of unknot diagrams D_n where the crossing number is a linear function of n but where any piecewise-linear spanning disc for the unknot must consist of at least exponentially many triangles.

To prove that there is a polynomial upper bound on Reidemeister moves, the author had to use both normal surfaces and arc presentations. As in Dynnikov's work (which is described in Section 2.3.8), one starts with an arc presentation of the unknot, and the aim is to apply exchange moves, cyclic permutations and destabilizations to reduce it to the trivial arc presentation. Dynnikov examined a spanning disc and its induced singular foliation, and showed that there is always a move that can be performed that results in a reduction in the number of singularities in the foliation or a reduction in the arc index. Using normal surface theory, it is not hard to show that, initially, the disc can be arranged to have at most exponentially many singularities. So, although this gives an upper bound on the number of moves, this bound is exponential. To obtain an improved estimate, one needs to go deeper into normal surface theory. It is possible to speak of two bits of normal surface as being *normally parallel*. The aim is to find many parallel pieces of the disc that can be moved all at the same time. In this way, one obtains, using a controlled number of exchange moves cyclic permutations and destabilizations, a substantial decrease in the complexity of the disc. This decrease is large enough that only polynomially many moves are required before the disc has been completely simplified, which implies that the arc presentation has become trivial.

2.4.4 An upper bound for each link type

The polynomial upper bound on Reidemeister moves described above has recently been generalized by the author as follows [58]. *For each link type K, there is a polynomial p_K such that any two diagrams of K with crossing number n and n' differ by a sequence of at most $p_K(n) + p_K(n')$ Reidemeister moves.* As a consequence, *for each link type K, the K-recognition problem is in NP.* The certificate is simply a sequence of Reidemeister moves taking the given diagram of K to some fixed diagram.

The proof uses and extends the techniques that were developed in the case of the unknot. In particular, both normal surfaces and arc presentations play a central role. In the case of the unknot, the spanning disc was simplified at each stage of the procedure. For a general knot K, one uses instead an entire hierarchy for the exterior of K. The spanning disc for the unknot could be simplified until its singular foliation contained just two singularities. This is not possible for the surfaces in a hierarchy. Instead, the goal is to simplify these surfaces until their number of singularities is bounded by a polynomial function of the initial arc index. Once this has been achieved, one considers a regular neighbourhood of this hierarchy together with a regular neighbourhood of the knot. This is a ball B, the complexity of which is, in some suitable sense, controlled. The ball inherits a handle structure by thickening the cell structure of the hierarchy, and K lies inside this handle structure in a way that depends only on the topology of the hierarchy. The proof proceeds by isotoping K through B until it lies within a single 0-handle and projects to some fixed diagram for K. Since B has controlled complexity, so too does the projection of K throughout this isotopy. Thus, one obtains a bound on the crossing number of the intermediate diagrams, and, with a bit more work, a polynomial bound on the number of Reidemeister moves used in this process.

2.4.5 Bounds on the crossing number of intermediate diagrams

In addition to bounding the number of Reidemeister moves required to pass between two link diagrams, one can also seek to bound the complexity of the diagrams in this sequence.

Problem 2.4. *Given two diagrams of a link, find a sequence of Reidemeister moves relating them where all diagrams in the sequence have small crossing number.*

This is, of course, related to Problem 2.3. First, an upper bound on the number of moves immediately gives an upper bound on the intermediate crossing numbers. This is because each Reidemeister move changes the crossing number by at most 2. Second, if one has an upper bound on the crossing number of each intermediate diagram, then one obtains an upper bound on the number of moves. This is because there are at most $(24)^{n+1}$ connected diagrams with crossing number at most n (see Theorem 6.5 in [9] for example). Moreover, in a shortest sequence of Reidemeister moves joining two diagrams, one never visits the same diagram twice. So, if each diagram in the sequence has crossing number at most n, say, then the length of the sequence is at most $(24)^{n+1}$.

However, one can sometimes establish a much better bound than this. As explained in Section 2.3.8, Dynnikov found, in the case of the unknot, a quadratic upper bound on the crossing number of intermediate diagrams.

2.5 Crossing number

The *crossing number* $c(K)$ of a knot or link K is defined to be the minimal number of crossings in any diagram for K.

2.5.1 Computing the crossing number

Crossing number is algorithmically computable. Indeed, this fact is closely related to the solubility of the equivalence problem for links. For example, the problem of determining whether a knot's crossing number is zero is clearly equivalent to determining whether it is the unknot. One can use a solution to the equivalence problem to determine a knot's crossing number in the following naive way. Starting with a knot diagram having n crossings, enumerate all diagrams with less than n crossings, and, for each, determine whether it is the same knot as the original one. The one with the smallest number of crossings clearly realizes the knot's crossing number. This approach is obviously not very efficient, but it seems unlikely that there is any quicker way of determining a link's crossing number in general. In fact, this method has been successfully used in practice to compile knot tables, by Hoste, Thistlethwaite and Weeks for example [32].

2.5.2 Determining the crossing number for certain link classes

Although there seems little hope of anything more than using a brute-force search to determine the crossing number of an arbitrary link, one can often determine the crossing number more efficiently in various interesting cases. For example, the crossing numbers of alternating links are completely understood, by work of Kauffman [40], Murasugi [72] and Thistlethwaite [97]. Their work, and a broader analysis of alternating links, is discussed in Section 2.7.1.

Torus links form another interesting collection, although they are of course more restricted. Murasugi [73] proved that the crossing number of the (p, q) torus link is $pq - \max\{p, q\}$.

2.5.3 The number of knot types

Although the crossing number is computable for any given knot, its behaviour is not well understood for infinite collections of knots. In particular, the following problem remains unresolved.

Problem 2.5. *If $N(c)$ is the number of knot types with crossing number c, determine the asymptotic behaviour of $N(c)$ as c tends to infinity.*

There is a more-or-less obvious exponential upper bound to $N(c)$. As explained in Section 2.4.5, the number of connected diagrams with crossing number c is at most $(24)^{c+1}$, and so $N(c) \leq (24)^{c+1}$.

One can also obtain an exponential lower bound on $N(c)$, by considering alternating knots. As described in Section 2.5.2, the crossing numbers of alternating knots are completely understood: a 'reduced' alternating diagram has minimal crossing number. (For an explanation of this terminology, see Section 2.7.1.) Moreover, the resolution of the Tait flyping conjecture by Menasco and Thistlethwaite [69] implies that one can determine exactly when two alternating diagrams represent the same knot type. So, enumerating the number of alternating knot types with crossing number c is nearly the same as enumerating the number of alternating diagrams. Very accurate asymptotics were established by Sundberg and Thistlethwaite [95], who showed that, if $N_{alt}(c)$ is the number of prime alternating link types with crossing number c, then

$$\lim_{c\to\infty} [N_{alt}(c)]^{1/c} = \frac{101 + \sqrt{21\,001}}{40}.$$

Using similar methods, Thistlethwaite [98] was also able to show that alternating links are increasingly scarce among all link types, as the crossing number tends to infinity. In particular, $N_{alt}(c)/N(c) \to 0$ as $c \to \infty$. But more precise asymptotics for $N(c)$ seem a long way from resolution.

2.5.4 The crossing number of composite knots

One of the most basic operations in knot theory is connected sum. Here, one starts with two oriented knots K_1 and K_2 in the 3-sphere. For each knot, one finds a 3-ball that intersects the knot in a properly embedded unknotted arc. One removes the interior of this ball. The result is, for each knot, a knotted arc in the complementary 3-ball. If one glues these two balls together, the result is the 3-sphere. One performs this gluing so that the two arcs join together to form a knot, and so that the orientations on the arcs patch together correctly. The resulting knot is the *connected sum* of K_1 and K_2, denoted by $K_1 \sharp K_2$.

As is evident from Figure 2.7, one can create a diagram D for $K_1 \sharp K_2$ starting from diagrams of K_1 and K_2. If one uses diagrams of K_1 and K_2 with minimal crossing number, one obtains a diagram D with crossing number $c(D)$ equal to $c(K_1) + c(K_2)$. Hence, $c(K_1 \sharp K_2) \leq c(K_1) + c(K_2)$. It seems 'obvious' that this should be an equality, but actually, it is far from clear, and this is in fact one of the most notorious unsolved problems in elementary knot theory.

Problem 2.6. *Must $c(K_1 \sharp K_2) = c(K_1) + c(K_2)$?*

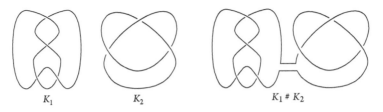

K_1 $\qquad\qquad$ K_2 $\qquad\qquad\qquad$ $K_1 \# K_2$

Figure 2.7 Connected sum.

This has been verified for certain knot classes, for example when K_1 and K_2 are both alternating [40, 72, 97] or when K_1 and K_2 are both torus knots [15]. The only general result in this direction is due to the author [52]: *For oriented knots K_1 and K_2,*

$$\frac{c(K_1) + c(K_2)}{152} \leq c(K_1 \sharp K_2) \leq c(K_1) + c(K_2).$$

We will give an overview of the proof, because it uses some unexpected methods that may be applicable to other problems in elementary knot theory.

There is a simple operation on knots and links that is closely related to the connected sum. This is the *disjoint union*. Here one starts with two links L_1 and L_2, each lying in the 3-sphere. One encloses each link in the interior of a ball and then identifies the boundaries of these balls, thereby forming a new link in the 3-sphere, which is denoted by $L_1 \sqcup L_2$.

Now, crossing number *does* behave well with respect to disjoint union. It is easy to prove that $c(L_1 \sqcup L_2) = c(L_1) + c(L_2)$. Again one has the upper bound $c(L_1 \sqcup L_2) \leq c(L_1) + c(L_2)$, because one can start with diagrams for L_1 and L_2 and use these to create a diagram for $L_1 \sqcup L_2$. To prove the inequality in the other direction, one must start with a diagram of $L_1 \sqcup L_2$ (which one may take to have minimal crossing number) and use this to create diagrams for L_1 and L_2. But this is straightforward: to form the diagram for L_1, say, one simply throws away the parts of the link projection that come from L_2. Thus, one obtains diagrams for L_1 and L_2, with the property that the sum of the number of crossings in these diagrams is at most the crossing number of the original diagram of $L_1 \sqcup L_2$. Thus, $c(L_1) + c(L_2) \leq c(L_1 \sqcup L_2)$.

One can immediately see from this the main difficulty in proving that $c(K_1) + c(K_2) \leq c(K_1 \sharp K_2)$. If one starts with a diagram D for $K_1 \sharp K_2$ with minimal crossing number, then there is no obvious way of forming diagrams for K_1 and K_2 from this. Although one might view a part of the curve $K_1 \sharp K_2$ as coming from K_1, say, this part is only an arc and so one needs to add in something extra to form a closed curve representing K_1. This may dramatically increase the number of crossings. This is evident when one considers the 2-sphere that is used to form the connected sum. If one performs an isotopy on $K_1 \sharp K_2$ so that it projects to give the diagram D, then the 2-sphere may be extremely distorted in 3-space. It is precisely an arc on this sphere that needs to be inserted to form a copy of K_1. The proof of the lower bound on $c(K_1 \sharp K_2)$ proceeds by controlling the complexity of this 2-sphere. In fact, one cuts $K_1 \sharp K_2$ at its two points of intersection with the sphere, and then one adds two arcs, one on either side of this sphere, to form a copy of $K_1 \sqcup K_2$. By controlling the sphere, one can control the number of new crossings this introduces. One thereby finds a diagram for $K_1 \sqcup K_2$ with crossing number at most $152\, c(D)$, which proves the theorem.

The way that the 2-sphere is controlled is via normal surface theory. One starts with a diagram D for $K_1 \sharp K_2$ with minimal crossing number. From this one can build a triangulation of the exterior M of $K_1 \sharp K_2$ in a reasonably natural way. The restriction of the 2-sphere to M is a properly embedded annulus A. Since A is essential, it can be isotoped into normal form. Hence, A intersects each tetrahedron in a collection of triangles and squares, as shown in Fig. 2.2. Within a tetrahedron, there are four types of triangle and three types of square. Between adjacent triangles or squares of the same type, there lies a product region. The union of these product regions forms an I-bundle $\mathcal{B}$ that is embedded in the exterior of A. This is known as the *parallelity bundle* for the exterior of A. The ∂I-bundle is the *horizontal boundary* of $\mathcal{B}$, denoted by $\partial_h \mathcal{B}$. Note that $\partial_h \mathcal{B}$ lies in the two copies of A in the exterior of A. The key part of the argument establishes that one can find properly embedded arcs in these two copies of A, each joining the two boundary components of the annulus, and avoiding $\partial_h \mathcal{B}$. The significance of this is that these arcs then can be made to intersect each tetrahedron of the triangulation in a restricted way. To see this, note that in each tetrahedron, all but at most 6 pieces in the exterior of a normal surface lie in $\mathcal{B}$. Since these arcs intersect each tetrahedron in a controlled way, one can bound from above the crossing number of their diagrammatic projection. Thus, we obtain a diagram for $K_1 \sharp K_2$, and, with some work, one can show that it has at most $152c(D)$ crossings.

2.5.5 The crossing number of satellite knots

A natural generalization of connected sum is the satellite knot construction (Fig. 2.8). A knot K is said to be a *satellite* of a non-trivial knot L if K lies in a regular neighbourhood $N(L)$ of L but does not lie within a 3-ball inside $N(L)$ and is not a core curve of $N(L)$. The knot L is called the *companion*.

The *wrapping number* of the satellite is the minimal number of intersections between K and a meridian disc for $N(L)$. It is always positive. Note that when K_1 and K_2 are non-trivial knots, then $K_1 \sharp K_2$ is a satellite of K_1 (and of K_2) with wrapping number 1.

The following problem remains famously unsolved.

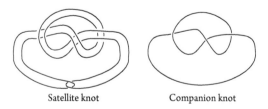

Satellite knot   Companion knot

Figure 2.8 A satellite knot.

Problem 2.7. *If K is a satellite of L, must c(K) be at least c(L)? More generally, must c(K)
be at least $w^2 c(L)$, where w is the wrapping number of the satellite?*

The motivation here is that there is a way of building a diagram for the satellite K
starting from a diagram for L. Each crossing of L becomes w^2 crossings for K. There may
be other crossings, and so, unlike in Problem 2.6, this is a conjectured inequality rather
than equality. One might be tempted to speculate that there is an equality relating $c(K)$
to the crossing number of L, the wrapping number and the crossing number of the knot
in $N(L)$, once this has been suitably defined. We do not do so here, although it is possible
that something along these lines may be true.

Unsurprisingly, the methods used to analyse the crossing number of composite knots
extend to the case of satellite knots, giving the following result of the author [54]: *If K is
a satellite of a knot L, then $c(K) \geq 10^{-13} c(L)$.*

In the proof, one starts with a diagram D of K with minimal crossing number, and the
goal is to create a diagram of L with crossing number at most $10^{13} c(D)$. One uses D to
construct a handle structure for the exterior of K. The satellite torus T can be placed in
'normal' form with respect to this handle structure. Cutting along T gives two pieces, one
a copy of the exterior of L, the other $N(L) - \text{int}(N(K))$. Each piece inherits a handle
structure. One then modifies this handle structure, primarily by replacing parts of the
parallelity bundle that are I-bundles over discs by 2-handles. These 2-handles may end
up lying in 3-space in a complicated way, but the remainder of the handle structure is
controlled. One then re-attaches $N(K)$ to $N(L) - \text{int}(N(K))$ to form a handle structure
of the solid torus $N(L)$. A theorem of the author [53] gives that in any handle structure of
the solid torus, there is a core curve that lies entirely within the 0-handles and 1-handles
and intersects each such handle in a restricted way. In this case, the core curve is a copy
of L, which then lies inside 3-space in a controlled way. Projecting this gives the required
diagram of L.

2.6 Crossing changes and unknotting number

An elementary operation that one can perform on a knot is to change a crossing in one of
its diagrams, as in Figure 2.9. This is strikingly difficult to analyse, and it leads to some of

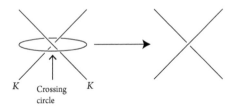

Figure 2.9 A crossing change.

the most challenging questions in the subject. An excellent survey of this topic can also be found in [90].

2.6.1 Computing unknotting number

The *unknotting number* $u(K)$ of a knot K is the minimal number of crossing changes that one can make to some diagram that turns it into the unknot. The difficulty of course is that one quantifies over all possible diagrams, and so although any diagram for K easily gives an upper bound on $u(K)$, it is rarely clear that this is the best possible. In fact, unknotting number is probably best understood in a manner that avoids diagrams: it is the minimal number of times that a knot must pass though itself in a one-parameter family of curves that starts with K and ends with the unknot.

Problem 2.8. *Is the unknotting number of a knot algorithmically computable? Is there even an algorithm to decide whether a given knot has unknotting number 1?*

The first of these questions seems a long way out of reach at present. It is more than just a theoretical question: there exist many explicit knots with unknown unknotting number (see [41] for example). As a result, many intriguing and clever methods have been developed that provide lower bounds on unknotting number [38, 39, 42, 60, 74, 76, 79, 94, 96]. These utilize a wide variety of techniques, including the Alexander module [74], the linking form on the double-branched cover [60, 94], 4-dimensional gauge theory [96] and, more recently, Heegaard Floer homology [76, 79]. A major piece of research in this area was the solution by Kronheimer and Mrowka [44] of Milnor's conjecture that *the unknotting number of the (p, q) torus knot is $(p-1)(q-1)/2$.*

However, with each new lower bound on unknotting number, there remain knots with unknotting number that cannot be determined. This will continue to be the case until Problem 2.8 is resolved.

A useful way of understanding and approaching this problem is via the use of crossing circles. Given some crossing in a diagram of a knot K, the associated *crossing circle* is a simple closed curve in the complement of K that encircles the crossing, as shown in Fig. 2.9. There are two crossing circles associated with each crossing (that are obtained from one another by rotating the diagram through $90°$). One usually focuses on the crossing circle that has zero linking number with K. This bounds a disc that intersects K in two points of opposite sign.

Crossing circles are useful for several reasons. First, the crossing change is achieved by ± 1 surgery along the crossing circle, and so one can draw on the well-developed theory of Dehn surgery. Second, it also permits crossings in different diagrams to be compared: the crossings are *equivalent* if their associated crossing circles are ambient isotopic in the complement of K. Clearly, changing equivalent crossings results in equivalent knots, and so this is a helpful concept. Some interesting problems can be phrased in terms of crossing circles.

Although it seems unlikely that we will have an algorithm to determine unknotting number in the near future, the question of whether knots with unknotting number 1 can be detected is a tantalizing one. Here, the situation seems considerably more promising. To answer this question, it would suffice to produce a finite list of potential crossing circles with the property that if K has unknotting number 1, then changing the crossing at one of these crossing circles would unknot the knot. For one could then check each of these crossing changes systematically to determine whether they did unknot the knot. (This would use one of the known solutions to the unknot recognition problem.) Thus, the following arises naturally.

Problem 2.9. *If a knot has unknotting number 1, are there only finitely many crossing circles, up to ambient isotopy, that can be used to unknot the knot?*

Much of what is known about knots with unknotting number one comes from sutured manifold theory. (See Section 2.3.5 for a very brief summary of some of this theory.) It is possible to show that if a knot K has unknotting number 1, then its exterior admits a taut sutured manifold hierarchy, *where the penultimate manifold in this decomposition is a solid torus neighbourhood of the crossing circle plus possibly some 3-balls*. In particular, this crossing circle is disjoint from some minimal-genus Seifert surface S for K. In fact, by pushing this argument further, Scharlemann and Thompson [89] were able to show that this crossing circle can be isotoped to lie in S. This suggests a route to determining whether a knot has unknotting number 1: find all minimal-genus Seifert surfaces for K and then search on each such surface for all possible crossing circles that unknot the knot. The first of these steps is possible, using normal surface theory, as long as K is not a satellite knot. However, there are infinitely many isotopy classes of curves on a surface with positive genus, and it is hard to see how to reduce this to a finite list of crossing circles on which to focus.

2.6.2 The unlinking and splitting numbers of a link

There is a natural analogue of unknotting number for links. The *unlinking number $u(L)$* of a link L is the minimal number of crossing changes required to turn it into the unlink. Many of the techniques that can be used to analyse unknotting number apply equally well to unlinking number. But they also often work in the following more general setting. The *splitting number $s(L)$* of a link L is the minimal number of crossing changes required to turn it into a split link.

Unlike the case of unknotting number 1, links with splitting number 1 or unlinking number 1 are known to be detectable, under fairly mild hypotheses. It is a recent theorem of the author [57] that *there is an algorithm to determine whether a hyperbolic link of at least 3 components has splitting number 1.*

This is proved as follows. Suppose that C is a crossing circle, such that changing this crossing turns a link L into a split link. Just as in Section 2.6.1, there is a sutured manifold hierarchy for the exterior of L where the penultimate manifold is a solid torus neighbourhood of C, plus some balls. The key part of the proof is to place this hierarchy into 'normal' form with respect to some initial triangulation of the exterior of L as in [50]. Thus, one obtains a finite computable list of possibilities for C. In particular, the version of Problem 2.9 in this setting is answered affirmatively. Moreover, each of these crossing changes can be analysed to determine whether it results in the unlink. So, under the above hypotheses, *there is also an algorithm to determine whether the link has unlinking number 1.*

2.6.3 Additivity of unknotting number

One can also ask how unknotting number behaves with respect to simple knot-theoretic operations. For example, the following is very far from resolution.

Problem 2.10. *Does $u(K_1 \sharp K_2) = u(K_1) + u(K_2)$ always hold? Less ambitiously, what about the inequality $u(K_1 \sharp K_2) \geq \max\{u(K_1), u(K_2)\}$?*

Scharlemann [87] proved that composite knots have unknotting number at least 2. But we cannot rule out the following bizarre possibility: two knots might each have unknotting number 100 but their connected sum has unknotting number 2!

2.6.4 Cosmetic crossing changes

A crossing circle is *nugatory* if it bounds a disc in the complement of the knot. Clearly changing a nugatory crossing does not alter the knot. Is this the only way?

Problem 2.11. *Suppose that a crossing change to a knot K does not alter its knot type. Let C be the associated crossing circle with zero linking number. Must C be nugatory?*

This is known when the knot K is the unknot, by work of Scharlemann and Thompson [89], which relied on Gabai's sutured manifold theory [21]. However, the general case seems to remain just out of reach.

2.7 Special classes of knots

Knots are usually specified by means of a diagram. Typically, this diagram reveals little immediate information about the knot. For example, a superficially complicated diagram may actually represent the unknot. However, there are certain types of diagram that force a knot to have much more structure. It is these that we discuss in this section.

2.7.1 Alternating knots

A diagram for a link is *alternating* if, as one travels along each component of the link, one meets 'over' and 'under' crossings alternately. An example of a non-alternating diagram, with successive 'over' crossings, is shown in Fig. 2.10. A link is *alternating* if it has an alternating diagram.

Some alternating diagrams can be immediately simplified as follows. We say that a diagram is *reducible* if, at some crossing, two of the incident regions are actually the same

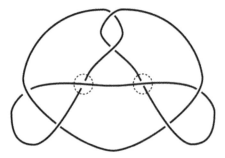

Figure 2.10 A non-alternating diagram (highlighted are successive 'over' crossings).

region of the diagram. When a diagram is reducible, there is an obvious modification that can be made to it, which reduces the crossing number by 1. When the original diagram is alternating, so too is the new diagram. So, when dealing with an alternating link, there is no loss of generality in assuming that its alternating diagram is not reducible. It is then known as *reduced*.

A great deal of information about a link is evident from an alternating diagram. Possibly the most striking illustration of this is the following result of Kauffman [40], Murasugi [72], and Thistlethwaite [97]: *If D is a reduced alternating diagram of a link L, then c(D) = c(L)*.

This was proved by means of the Jones polynomial. This link invariant is defined in terms of some diagram. It is therefore unsurprising that the 'complexity' of the polynomial should be bounded above by the 'complexity' of some diagram. In fact, the *width* of the polynomial, which is the difference in degree between the highest-order term and the lowest-order term, is always at most the crossing number of any diagram. But, Kauffman, Murasugi and Thistlethwaite proved that this inequality is an equality for a reduced alternating diagram D. This implies the above result, because the width of the Jones polynomial is then the crossing number of D, and the link cannot have a diagram with fewer crossings, because this would force the width of the polynomial to be at most this number.

A consequence of this theorem is that there is a very rapid and simple algorithm to decide whether an alternating diagram represents the unknot: *an alternating diagram of the unknot either has no crossings or is reducible*. In fact, this was already known, owing to earlier techniques of Menasco [68], which relied on a very precise analysis of surfaces in the exterior of an alternating link. This works particularly well when the surface has non-negative Euler characteristic. So, by examining spheres in the complement of the link, one can also detect whether an alternating link is split. Menasco showed that *an alternating link is split if and only if its alternating diagram is disconnected*. Similarly, *an alternating knot is prime if and only if a reduced alternating diagram is prime*. Here, a connected diagram is *prime* if and only if any simple closed curve in the plane that intersects the link projection transversely in two points away from the crossings divides the plane into two components, one of which contains no crossings.

In Menasco's proof, he constructs, using the diagram, a special position for the link in $\mathbb{R}^3$. The diagram is viewed as lying in the horizontal plane $\mathbb{R}^2 \times \{0\}$. The link mostly

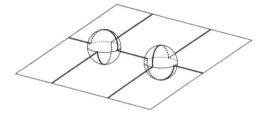

Figure 2.11 Balls near crossings.

lies in this plane, except near each crossing, where two sub-arcs of the link take detours above and below the plane of the diagram. More precisely, one places a small 3-ball at each crossing with centre lying in the plane of the diagram, and the two arcs involved in the crossing follow the boundary of this ball. (See Fig. 2.11.) Let $\mathbb{R}^3_+$ denote the set of points lying above the union of the plane and the balls, and define $\mathbb{R}^3_-$ similarly. Menasco showed that a closed incompressible surface in the link exterior may be arranged so that it intersects each of these balls in a collection of parallel 'saddles', and intersects $\mathbb{R}^3_+$ and $\mathbb{R}^3_-$ in a collection of discs. The boundary of these discs in $\mathbb{R}^3_+$ is a collection of simple closed curves, which can be viewed as lying within the diagram. A careful analysis of these curves (particularly, an innermost one in the plane), together with the condition that the diagram is connected and alternating, implies that the surface S has a *meridional compression disc*. This is a disc D in $\mathbb{R}^3$ such that $D \cap S = \partial D$ and $D \cap K$ is a single point in the interior of D. The existence of such a disc is impossible for a splitting sphere, for example, and this can be used to establish Menasco's theorem about split alternating links.

A significantly deeper version of this analysis was used to prove the following theorem of Menasco and Thistlethwaite [69]: *Any two reduced alternating diagrams of a link differ by a sequence of flypes* (as shown in Fig. 2.12). This was Tait's flyping conjecture. The proof of this is very complex, and so we can only give a very brief overview. Each alternating diagram determines two chequerboard surfaces. In Menasco and Thistlethwaite's proof, they examine the surfaces arising from one diagram and how they interact with other diagram. A careful analysis, together with a surprising intervention of the Jones polynomial at one point, establishes the existence of a flype that, in some sense, simplifies these surfaces, and so the theorem is proved by induction.

Although alternating links are now very well understood, there are some invariants and properties of these links that are worthy of further analysis. Some of these are given in the following problem.

Problem 2.12. *Can one detect (or possibly just estimate) the following invariants of a knot, given an alternating diagram: its bridge number; its tunnel number; its unknotting number?* [3]

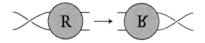

Figure 2.12 Flype.

One may also ask for estimates for more sophisticated invariants. Indeed, in [51], the author has given a way of reading off the hyperbolic volume of an alternating link's complement, up to a bounded multiplicative error. While such problems are interesting, they falls outside the remit of this survey, which is focused on much more elementary questions.

2.7.2 Positive knots and positive braids

Given the manifest success in understanding the topological properties of alternating knots, it is natural to consider other possible classes of knot diagrams. A reasonable way of assessing whether it is a good class of diagrams is to ask whether it satisfies the minimal requirement that a diagram specifies the unknot if and only if it 'obviously' does so. Another class of diagrams satisfying this requirement is the class of positive diagrams.

Positivity is defined for diagrams of oriented links. Positive and negative crossings are shown in Fig. 2.13. A diagram is said to be *positive* if all its crossings are positive.

Cromwell [11] showed that a reduced positive diagram never represents the unknot. In fact, one can read off the genus of a positive knot from its positive diagram. This result was significantly generalized by Kronheimer and Mrowka [44], who showed that the smooth 4-ball genus of a positive knot can be read off from its positive diagram. Indeed, it is equal to the knot's genus.

However, positive knots are not totally well behaved. For example, Stoimenow [93] found a positive knot that has no positive diagram with minimal crossing number. So, it seems unlikely that one can read off the crossing number of a positive knot in the way that one can for alternating knots. Nevertheless, they form an interesting class that is worthy of further investigation. In particular, it would be instructive to determine other topological and/or elementary invariants of positive knots. For example, is there a simple relationship between two positive diagrams of the same knot?

One can also consider, more specifically, closed positive braids. These appear to have even nicer properties than general positive knots. For example, they are fibred. They may be more amenable to analysis.

2.7.3 Other interesting classes of diagram

There are several other related classes of diagrams. *Homogeneous* diagrams form a common generalization of positive diagrams and alternating diagrams. Cromwell [11] in fact established his formula for knot genus not just for positive knots but for homogeneous ones.

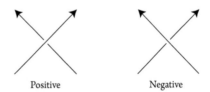

Positive   Negative

Figure 2.13 Positive and negative crossings.

Adequate diagrams form another interesting class. These are defined by considering 'resolutions' for each crossing, which remove the crossing. Each crossing may be removed in one of two ways: a 'plus' or a 'minus' resolution. If one resolves every crossing, the result is a collection of planar curves. At each crossing, the two new arcs may belong to the same simple curve or two distinct ones. A diagram is *plus-adequate* if, when all the crossings are resolved in a 'plus' way, each resulting simple closed curve runs through the remnants of each crossing at most once. *Minus-adequacy* is defined similarly using the 'minus' resolution at each crossing. A diagram is *adequate* if it is both plus-adequate and minus-adequate. Adequate knots have well-behaved Jones polynomials. Lickorish and Thistlethwaite [61] used this to prove that *if a knot has a reduced adequate diagram with at least one crossing, then it is not the unknot.* But adequacy has more topological applications, as shown in the recent monograph of Futer, Kalfagianni and Purcell [20]. Adequate links surely merit further investigation.

2.8 Epilogue

The theory of 3-manifolds has seen remarkable progress over the past few years, with the solution of many key conjectures, for example, the Poincaré Conjecture [81–83] and the Virtually Haken Conjecture [2], to name just two. One might be led to the conclusion that the related field of knot theory might be devoid of interesting unsolved problems. The purpose of the present chapter has been to emphasize that this is not the case. There remain many fundamental and interesting challenges ahead.

Notes

1. Since the first version of this chapter was produced, two new solutions to the related problem of determining whether two closed 3-manifolds are homeomorphic have been given by Scott and Short [91] and by Kuperberg [49]. Both rely on the solution to the Geometrization Conjecture, as in Section 2.2.2.
2. This approach has now been successfully completed by the author [56], giving the first full unconditional proof that unknot recognition is in co-NP.
3. An algorithm to determine whether an alternating knot has unknotting number 1 has recently been given by McCoy [67].

References

[1] I. Agol, *Knot genus is NP*, conference presentation (2002).
[2] I. Agol, *The Virtual Haken Conjecture* (with appendix by I. Agol, D. Groves and J. Manning), Doc. Math. **18** (2013) 1045–1087.
[3] I. Agol, J. Hass and W. Thurston, *The computational complexity of knot genus and spanning area*, Trans. Am. Math. Soc. **358** (2006) 3821–3850.
[4] D. Bennequin, *Entrelacements et équations de Pfaff*, Astérisque **107–108** (1983) 87–161.

[5] J. Birman and T. Brendle, *Braids: a survey*, in W. Menasco and M. Thistlethwaite, eds., *Handbook of Knot Theory* (Elsevier, 2005), pp. 19–103.

[6] J. Birman and M. Hirsch, *A new algorithm for recognizing the unknot*, Geom. Topol. **2** (1998) 178–220.

[7] J. Birman, W. Menasco, *Studying links via closed braids V: Closed braid representatives of the unlink*, Trans. Am. Math. Soc. **329** (1992) 585–606.

[8] M. Brittenham, *Essential laminations and Haken normal form*, Pac. J. Math. **168** (1995) 217–234.

[9] A. Coward and M. Lackenby, *An upper bound on Reidemeister moves*, Am. J. Math. **136** (2014) 1023–1066.

[10] D. Coulson, O. Goodman, C. Hodgson and W. Neumann, *Snap*, http://www.ms.unimelb.edu.au/~snap/.

[11] P. Cromwell, *Homogeneous links.* J. Lond. Math. Soc. (2) **39** (1989) 535–552.

[12] P. Cromwell, *Embedding knots and links in an open book I: basic properties*, Topol. Applic. **64** (1995) 37–58.

[13] M. Culler and P. Shalen, *Bounded, separating, incompressible surfaces in knot manifolds*, Invent. Math. **75** (1984) 537–545.

[14] F. Dahmani and D. Groves, *The isomorphism problem for toral relatively hyperbolic groups*, Publ. Math. Inst. Hautes Études Sci. **107** (2008), 211–290.

[15] Y. Diao, *The additivity of crossing numbers*, J. Knot Theory Ramif. **13** (2004) 857–866.

[16] S. Donaldson, *An application of gauge theory to four-dimensional topology*, J. Differential Geom. **18** (183) 279–315.

[17] I. Dynnikov, *Arc-presentations of links: monotonic simplification*, Fund. Math. **190** (2006), 29–76.

[18] D. Epstein and R. Penner, *Euclidean decompositions of noncompact hyperbolic manifolds*, J. Differential Geom. **27** (1988) 67–80.

[19] A. Floer, *An instanton-invariant for 3-manifolds*, Commun. Math. Phys. **118** (1988) 215–240.

[20] D. Futer, E. Kalfagianni and J. Purcell, *Guts of Surfaces and the Colored Jones Polynomial*, Lecture Notes in Mathematics, No. 2069 (Springer-Verlag, 2013).

[21] D. Gabai, *Foliations and the topology of 3-manifolds*, J. Differential Geom. **18** (1983) 445–503.

[22] D. Gabai, *Essential laminations and Kneser normal form*, J. Differential Geom. **53** (1999) 517–574.

[23] C. Gordon and J. Luecke, *Knots are determined by their complements*, J. Am. Math. Soc. **2** (1989) 371–415.

[24] W. Haken, *Theorie der Normalflächen*, Acta Math. **105** (1961) 245–375.

[25] W. Haken, *Some results on surfaces in 3-manifolds*, in P. J. Hilton, ed., *Studies in Modern Topology* (Mathematical Association of America/Prentice-Hall, 1968), pp. 39–98.

[26] J. Hass, J. Lagarias, *The number of Reidemeister moves needed for unknotting*, J. Am. Math. Soc. **14** (2001) 399–428

[27] J. Hass, J. Lagarias and N. Pippenger, *The computational complexity of knot and link problems*, J. ACM **46** (1999) 185–211.

[28] J. Hass and T. Nowik, *Unknot diagrams requiring a quadratic number of Reidemeister moves to untangle*, Discrete Comput. Geom. **44** (2010) 91–95.

[29] J. Hass, J. Snoeyink and W. Thurston, *The size of spanning disks for polygonal curves*, Discrete Comput. Geom. **29** (2003) 1–17.

[30] G. Hemion, *On the classification of homeomorphisms of 2-manifolds and the classification of 3-manifolds*, Acta Math. **142** (1979) 123–155.

[31] J. Hoste and M. Thistlethwaite, *Knotscape*, http://www.math.utk.edu/~morwen/knotscape. html.

[32] J. Hoste, M. Thistlethwaite and J. Weeks, *The first 1,701,936 knots*, Math. Intelligencer **20** (1998) 33–48.

[33] W. Jaco and P. Shalen, *Seifert fibered spaces in 3-manifolds*, Mem. Am. Math. Soc. **21** (1979) No. 220.

[34] W. Jaco and J. Tollefson, *Algorithms for the complete decomposition of a closed 3-manifold*, Illinois J. Math. **39** (1995) 358–406.

[35] F. Jaeger, D. Vertigana and D. Welsh, *On the computational complexity of the Jones and Tutte polynomials*, Math. Proc. Camb. Phil. Soc. **108** (1990) 35–53.

[36] K. Johannson, *Homotopy Equivalences of 3-Manifolds with Boundaries*, Lecture Notes in Mathematics, No. 761 (Springer-Verlag, 1979).

[37] V. Jones, *A polynomial invariant for knots via von Neumann algebras*, Bull. Am. Math. Soc. (N.S.) **12** (1985) 103–111.

[38] T. Kawamura, *The unknotting numbers of 10_{139} and 10_{152} are 4*, Osaka J. Math. **35** (1998) 539–546.

[39] T. Kanenobu and H. Murakami, *Two-bridge knots with unknotting number one*, Proc. Am. Math. Soc. **98** (1986) 499–502.

[40] L. Kauffman, *State models and the Jones polynomial*, Topology **26** (1987) 395–407.

[41] R. Kirby, *Problems in low dimensional manifold theory*, Proc. Symp. Pure Math. **32**(2) (1978) 273–312.

[42] T. Kobayashi, *Minimal genus Seifert surface for unknotting number 1 knots*, Kobe J. Math. **6** (1989) 53–62.

[43] P. Koiran, *Hilbert's Nullstellensatz is in the polynomial hierarchy*, J. Complexity **12** (1996) 273–286.

[44] P. Kronheimer and T. Mrowka, *Gauge theory for embedded surfaces. I*, Topology **32** (1993) 773–826.

[45] P. Kronheimer and T. Mrowka, *Dehn surgery, the fundamental group and SU(2)*, Math. Res. Lett. **11** (2004) 741–754.

[46] P. Kronheimer and T. Mrowka, *Monopoles and Three-Manifolds* (Cambridge University Press, 2007).

[47] P. Kronheimer and T. Mrowka, *Khovanov homology is an unknot-detector*, Publ. Math. Inst. Hautes Études Sci. **113** (2011) 97–208.

[48] G. Kuperberg, *Knottedness is in NP, modulo GRH*, Adv. Math. **256** (2014) 493–506.

[49] G. Kuperberg, *Algorithmic homeomorphism of 3-manifolds as a corollary of geometrization*, arXiv:1508.06720 [math.GT].

[50] M. Lackenby, *Exceptional surgery curves in triangulated 3-manifolds*, Pac. J. Math. **210** (2003) 101–163.

[51] M. Lackenby, *The volume of hyperbolic alternating link complements*, Proc. Lond. Math. Soc. **88** (2004) 204–224.

[52] M. Lackenby, *The crossing number of composite knots*, J. Topol. **2** (2009) 747–768.

[53] M. Lackenby, *Core curves of triangulated solid tori*, Trans. Am. Math. Soc. **366** (2014) 6027–6050; arXiv:1106.2934 [math.GT].

[54] M. Lackenby, *The crossing number of satellite knots*, Algebr. Geom. Topol. **14** (2014) 2379–2409; arXiv:1106.3095 [math.GT].

[55] M. Lackenby, *A polynomial upper bound on Reidemeister moves*, Ann. Math. **182** (2015) 491–564.

[56] M. Lackenby, *The efficient certification of knottedness and Thurston norm*, arXiv:1604.00290.

[57] M. Lackenby, *Links with splitting number one*, In preparation.

[58] M. Lackenby, *A polynomial upper bound on Reidemeister moves for each link type*, In preparation.

[59] J. Lagarias and A. Odlyzko, *Effective versions of the Chebotarev density theorem*, in A. Frohlich, ed., *Algebraic Number Fields: L-functions and Galois Properties* (Academic Press, 1977), pp. 409–464.

[60] W. B. R. Lickorish, *The unknotting number of a classical knot*. Contemp. Math. **44** (1985) 117–121.

[61] W. B. R. Lickorish and M. Thistlethwaite, *Some links with nontrivial polynomials and their crossing-numbers*, Comment. Math. Helv. **63** (1988) 527–539.

[62] A. Mal'cev, *On isomorphic matrix representations of infinite groups*, Mat. Sb. **8**(50) (1940), 405–422.

[63] J. Manning, *Algorithmic detection and description of hyperbolic structures on closed 3-manifolds with solvable word problem*, Geom. Topol. **6** (2002), 1–25.

[64] C. Manolescu, P. Ozsváth and S. Sarkar, *A combinatorial description of knot Floer homology*, Ann. Math. **169** (2009) 633–660.

[65] A. Markov, *The unsolvability of the homeomorphy problem*, Dokl. Akad. Nauk SSSR **121** (1958) 218–220.

[66] S. Matveev, *Algorithmic Topology and Classification of 3-Manifolds*, 2nd edn (Springer-Verlag, 2007).

[67] D. McCoy, *Alternating knots with unknotting number one*, arXiv:1312.1278 [math.GT].

[68] W. Menasco, *Closed incompressible surfaces in alternating knot and link complements*, Topology **23** (1984) 37–44.

[69] W. Menasco and M. Thistlethwaite, *The classification of alternating links*, Ann. Math. **138** (1993) 113–171.

[70] A. Mijatovic, *Simplifying triangulations of S^3*, Pac.J. Math. **208** (2003) 291–324.

[71] A. Mijatovic, *Simplicial structures of knot complements*, Math. Res. Lett. **12** (2005) 843–856.

[72] K. Murasugi, *Jones polynomials and classical conjectures in knot theory*, Topology **26** (1987) 187–194.

[73] K. Murasugi, *On the braid index of alternating links*, Trans. Am. Math. Soc. **326** (1991) 237–260.

[74] Y. Nakanishi, *A note on unknotting number*, Math. Sem. Notes Kobe Univ. **9** (1981) 99–108.

[75] W. Neumann and G. Swarup, *Canonical decompositions of 3-manifolds*, Geom. Topol. **1** (1997) 21–40.

[76] B. Owens, *Unknotting information from Heegaard Floer homology*, Adv. Math. **217** (2008) 2353–2376.

[77] P. Ozsváth and Z. Szabó, *Holomorphic disks and topological invariants for closed three-manifolds*, Ann. Math. **159** (2004) 1027–1158.

[78] P. Ozsváth and Z. Szabó, *Holomorphic disks and genus bounds*, Geom. Topol. **8** (2004) 311–334.

[79] P. Ozsváth and Z. Szabó, *Knots with unknotting number one and Heegaard Floer homology*, Topology **44** (2005) 705–745.

[80] U. Pachner, *P.L. homeomorphic manifolds are equivalent by elementary shellings*, Eur. J. Combin. **12** (1991) 129–145.

[81] G. Perelman, *The entropy formula for the Ricci flow and its geometric applications*, arXiv:math/0211159 [math.DG].

[82] G. Perelman, *Ricci flow with surgery on three-manifolds*, arxiv:math/0303109 [math.DG].

[83] G. Perelman, *Finite extinction time for the solutions to the Ricci flow on certain three-manifolds*, arXiv:math/0307245 [math.DG].

[84] V. Pratt, *Every prime has a succinct certificate*, SIAM J. Comp. **4** (1975) 214–220.

[85] K. Reidemeister, *Knotten und Gruppen*, Abh. Math. Sem. Univ. Hamburg **5** (1927) 7–23.

[86] S. Sarkar and J. Wang, *An algorithm for computing some Heegaard Floer homologies*, Ann. Math. **171** (2010) 1213–1236.

[87] M. Scharlemann, *Unknotting number one knots are prime*, Invent. Math. **82** (1985) 37–55.

[88] M. Scharlemann, *Sutured manifolds and generalized Thurston norms*, J. Differential Geom. **29** (1989) 557–614.

[89] M. Scharlemann and A. Thompson, *Link genus and the Conway moves*, Comment. Math. Helv. **64** (1989) 527–535.

[90] M. Scharlemann, *Crossing changes*, Chaos Solitons Fractals **9** (1998) 693–704.

[91] P. Scott, H. Short, *The homeomorphism problem for closed 3-manifolds*, Algebr. Geom. Topol. **14** (2014) 2431–2444.

[92] Z. Sela, *The isomorphism problem for hyperbolic groups. I*, Ann. Math. **141** (1995) 217–283.

[93] A. Stoimenow, *On the crossing number of positive knots and braids and braid index criteria of Jones and Morton–Williams–Franks*, Trans. Am. Math. Soc. **354** (2002), 3927–3954.

[94] A. Stoimenow, *Polynomial values, the linking form and unknotting numbers*, Math. Res. Lett. **11** (2004) 755–769.

[95] C. Sundberg and M. Thistlethwaite, *The rate of growth of the number of prime alternating links and tangles*, Pac. J. Math. **182** (1998) 329–358.

[96] T. Tanaka, *Unknotting numbers of quasipositive knots*, Topol. Applic. **88** (1998) 239–246.

[97] M. Thistlethwaite, *A spanning tree expansion of the Jones polynomial*, Topology **26** (1987) 297–309.

[98] M. Thistlethwaite, *On the structure and scarcity of alternating links and tangles*, J. Knot Theory Ramif. **7** (1998) 981–1004.

[99] A. Thompson, *Thin position and the recognition problem for S^3*, Math. Res. Lett. **1** (1994) 613–630.

[100] W. Thurston, *Three-dimensional manifolds, Kleinian groups and hyperbolic geometry*, Bull. Am. Math. Soc. (N.S.) **6** (1982) 357–381.

[101] W. Thurston, *A norm for the homology of 3-manifolds*, Mem. Am. Math. Soc. **59** (1986) i–vi and 99–130.

[102] J. Weeks, *Snappea*, http://www.geometrygames.org/SnapPea/.

[103] J. Weeks, *Convex hulls and isometries of cusped hyperbolic 3-manifolds*, Topol. Applic. **52** (1993) 127–149.

[104] P. Weinberger, *Finding the number of factors of a polynomial*, J. Algorithms **5** (1984) 180–186.

[105] E. Witten, *Quantum field theory and the Jones polynomial*, Commun. Math. Phys. **121** (1989) 351–399.

[106] E. Witten, *Monopoles and four-manifolds*, Math. Res. Lett. **1** (1994) 769–796.

3 Cube Complexes, Subgroups of Mapping Class Groups and Nilpotent Genus

MARTIN R. BRIDSON

3.1 Introduction

These notes are based on my lecture at PCMI in July 2012. They are structured around two sets of results, one concerning groups of automorphisms of surfaces and the other concerning the nilpotent genus of groups. The first set of results exemplifies the theme that even the nicest of groups can harbour a diverse array of complicated finitely presented subgroups: we shall see that the finitely presented subgroups of the mapping class groups of surfaces of finite type can be much wilder than had been previously recognized. The second set of results fits into the quest to understand which properties of a finitely generated group can be detected by examining the group's finite and nilpotent quotients and which cannot.

These two topics appear to have little in common and neither has any obvious connection to the study of non-positively curved cube complexes—they are chosen for exactly these reasons. My purpose is to describe the resolution of various longstanding problems in a way that emphasizes the broad applicability of a certain template for constructing interesting examples of finitely presented groups. This template, described in Section 3.8, can be applied in many other contexts. It refines a construction that I articulated in [17], with improvements based on recent advances in the understanding of right-angled Artin groups (RAAGs) and non-positively curved cube complexes (particularly the virtually special cube complexes of Haglund and Wise).

The new results concerning subgroups of mapping class groups are from [16] while the new results concerning nilpotent genera of groups are from [29] (which is part of a

Lectures on Geometry. Edward Witten, Marc Lackenby, Martin R. Bridson, Helmut Hofer and Rahul Pandharipande.
© Oxford University Press 2017. Published 2017 by Oxford University Press.

wider project with Alan Reid from the University of Texas). Two of the results that we shall discuss are the following.

Theorem 3.1 (from [16]). *If the genus of a surface S is sufficiently large, then the isomorphism problem for the finitely presented subgroups of the mapping class group Mod(S) is unsolvable.*

Theorem 3.2 (from [29]). *There exist pairs of finitely generated, residually torsion-free nilpotent groups $N \hookrightarrow \Gamma$ so that N has the same finite and nilpotent quotients as Γ but Γ is finitely presented while $H_2(N, \mathbb{Q})$ is infinite-dimensional.*

Several of the results that we shall discuss concern decision problems for groups. I shall assume that the reader is familiar with the basic vocabulary associated with such problems, and recommend [56] as a pleasant introduction to the subject.

3.2 Subgroups of mapping class groups

Throughout, S will denote a compact, connected, orientable surface, which is allowed to have non-empty boundary. Most of the results that we will discuss remain valid for compact surfaces with finitely many punctures. The *mapping class group* of S will be denoted by[1] Mod(S); this is the group of isotopy classes of orientation-preserving homeomorphisms $S \to S$, where maps and isotopies are required to fix the boundary pointwise. I shall assume that the reader is familiar with the basic ideas and vocabulary concerning mapping class groups, as described in [42] and [50] for example.

3.2.1 The first subgroups

3.2.1.1 *Cyclic subgroups*

Nielsen–Thurston theory describes the individual elements of Mod(S): an element $\phi \in$ Mod(S) is *reducible* if it is the class of a homeomorphism that leaves invariant a non-empty collection of homotopically essential circles on S, none of which is homotopic to a boundary component; Thurston proved that the irreducible elements ψ of infinite order are *pseudo-Anosov*, which implies in particular that if c is a loop that is not homotopic into ∂S, then, in any fixed metric, the length of the shortest loop in the homotopy class $\psi^n[c]$ grows exponentially with $|n|$.

Nielsen proved that the elements of finite order in Mod(S) are precisely those mapping classes that contain a diffeomorphism f such that $f^d = \text{id}_S$ for some d.

3.2.1.2 *Finite subgroups*

Kerckhoff's resolution of the Nielsen Realization Problem [51] shows that every finite subgroup $G < \text{Mod}(S)$ arises as a group of isometries of a metric of constant curvature on S; equivalently, G has a fixed point in the natural action of Mod(S) on the Teichmüller space $\mathcal{T}(S)$. The action of Mod(S) on $\mathcal{T}(S)$ is proper and there is an equivariant retraction onto a spine where the action of Mod(S) is cocompact [59]. It follows that there are only *finitely many conjugacy classes of finite subgroups in* $Mod(S)$; see [20]. This does not remain true if one replaces $Mod(S)$ by a finitely presented subgroup (Theorem 3.6).

3.2.1.3 Abelian subgroups

The Dehn twists in disjoint curves on S have infinite order and commute. On a closed surface of genus g, the maximum number of disjoint, non-homotopic, essential simple closed curves that one can fit is $3g - 3$; if S has b boundary components, then one can add a further b curves parallel to the boundary components. Thus we obtain free abelian groups of rank $3g - 3 + b$ generated by Dehn twists. Birman, Lubotzky and McCarthy [14] proved that every abelian subgroup of $Mod(S)$ is finitely generated and has rank no greater than $3g - 3 + b$ (see also [50]).

3.2.1.4 Free subgroups

These abound in $Mod(S)$; indeed, Ivanov[2] [50] and McCarthy [54] proved that mapping class groups satisfy a *Tits alternative*: if a subgroup $G < Mod(S)$ is not virtually abelian, then it contains a free subgroup. More explicitly, if $\phi, \psi \in Mod(S)$ are pseudo-Anosov, then, for some $n > 0$, the subgroup $\langle \phi^n, \psi^n \rangle$ is free (and cyclic only if ϕ and ψ share an axis in Teichmüller space).

A more involved argument due to Dahmani, Guirardel and Osin [36] shows that if $\psi \in Mod(S)$ is pseudo-Anosov, then, for suitable $m > 0$, the normal closure $\langle\langle \psi^m \rangle\rangle$ is a free group freely generated by the conjugates of ψ^m.

3.2.1.5 Direct products

If S' and S'' are disjoint subsurfaces of S, then, by extending diffeomorphisms to be the identity on the complement (which is assumed to have no annular components), we see that $Mod(S') \times Mod(S'')$ is a subgroup of $Mod(S)$.

3.2.2 Wreath products

The wreath product $A \wr B$ of groups is the semidirect product $B \ltimes \oplus_{b \in B} A_b$, where the A_b are isomorphic copies of A permuted by left translation.

Proposition 3.3. *If S is a compact surface with non-empty boundary and G is a finite group, then there is a closed surface S_g and a monomorphism $Mod(S) \wr G \to Mod(S_g)$.*

To prove this, one first reduces to the case where S has one boundary component. Then one takes a closed surface on which G acts effectively, deletes a family of open discs centred at the points of a free G-orbit and attaches a copy of S to each of the resulting boundary circles, extending homeomorphisms of each copy of S by the identity on the complement. (See [18] for details.)

For a group H with finite-index $K \lhd H$, there is a standard embedding $H \hookrightarrow K \wr (H/K)$.

Corollary 3.4. *If a group H has a subgroup of finite index that embeds in the mapping class group of a compact surface with boundary, then H embeds in the mapping class group of a closed surface.*

Remark 3.5. There is considerable flexibility in the above construction, but one cannot hope to embed H in $\text{Mod}(S)$ for *all* surfaces of sufficiently high genus. Indeed, there are constraints even for finite groups [53].

3.2.3 Non-subgroups

3.2.3.1 *Reducibility and its consequences*

The reduction theory of Ivanov [50] allows one to prove all manner of results concerning the subgroups of mapping class groups by induction on the complexity of the surface. To explain this, we first extend the definition of reducibility to subgroups: $G < \text{Mod}(S)$ is *reducible* if it leaves invariant a non-empty collection of homotopically essential circles on S, none of which is homotopic to a boundary component. Ivanov [50] proves that if $G < \text{Mod}(S)$ is infinite and irreducible, then it contains a pseudo-Anosov element.

There are subgroups of finite index $P < \text{Mod}(S)$ that consist entirely of *pure* automorphisms: $[\phi]$ is pure if, on the complement of a tubular neighbourhood of a set of disjoint curves that ϕ fixes, the restriction of ϕ to each component is either trivial or pseudo-Anosov. The kernel of the action of $\text{Mod}(S)$ on $H_1(S, \mathbb{Z}_3)$ is pure, for example [50, p.4]. With this fact in hand, it follows from Ivanov's theorem that, after passing to a subgroup of finite index, any $G < \text{Mod}(S)$ fits into a short exact sequence where the kernel (which is central) is a free virtually abelian group generated by multitwists in the reducing curves for G and the quotient is a direct product of subgroups of mapping class groups of smaller surfaces, each containing a pseudo-Anosov on that surface.

From there, it is not difficult to prove, for example, that every amenable subgroup is abelian (cf. [14, 50]). And, using [36], one sees that $\text{Mod}(S)$ does not contain infinite images of groups that are $\mathbb{Z}$-averse in the sense of [30], from which it follows that all homomorphisms to $\text{Mod}(S)$ from irreducible lattices in higher-rank semisimple Lie groups have finite image (cf. [43]).

Other obstructions to embedding come from global properties of $\text{Mod}(S)$ that are inherited by subgroups; for example, if $H < \text{Mod}(S)$, then H must be residually finite [44], and all cyclic subgroups of H must be quasi-isometrically embedded [41].

3.2.4 Subgroups that are not finitely presented

We are concerned here almost entirely with subgroups that are finitely presented, or at least finitely generated. But it would be remiss of me not to mention that $\text{Mod}(S)$ has a host of natural subgroups where these finiteness conditions may fail. Foremost among these is the *Torelli group*, which is the kernel of the action of $\text{Mod}(S)$ on $H_1(S, \mathbb{Z})$. This is finitely generated if the genus of S is at least 3, but it is not known if it is finitely presented.

The kernel of the action Mod(S) on $\pi_1(S)$ modulo any later term of the lower central series is not even finitely generated. Connecting to the second part of this chapter, we note that the Torelli group is *residually torsion-free nilpotent* [5].

3.2.5 On the difficulty of identifying finitely presented subgroups

For the most part, the heart of geometric and combinatorial group theory lies with the study of *finitely presented* groups, but this restriction brings with it real challenges. In any context, it is easy to find finitely generated subgroups of a given group G: one can simply take a finite subset $S \subset G$ and consider $\langle S \rangle$. But typically there will be no algorithm to decide which finite subsets of G generate finitely presentable subgroups (this is the case in a direct product of free groups already). Moreover, even if one is given information that guarantees that $\langle S \rangle$ has a finite presentation, there is no general procedure that will produce such a presentation, even when the ambient group is as benign as $G = \mathrm{GL}(n, \mathbb{Z})$; see [30].

3.2.6 Direct products of free groups

If $H_1, H_2 < \mathrm{Mod}(S)$ are supported on disjoint subsurfaces of S, then they commute. One can embed g disjoint one-holed tori in a surface of genus g, and the mapping class group of a one-holed torus is a central extension of $\mathrm{SL}(2, \mathbb{Z})$, which contains non-abelian free groups. Thus, if S has genus g (and any number of boundary components), then Mod(S) contains the direct product D of g non-abelian free groups. In D, what (finitely generated or finitely presented) subgroups might we find? The answer to this question leads us in the main direction of this part of the chapter. But first we pause to record a consequence of Corollary 3.4.

Theorem 3.6. *There exist closed surfaces S and finitely presented subgroups $H < \mathrm{Mod}(S)$ such that H has infinitely many conjugacy classes of elements of finite order.*

Proof. The direct product Γ of n copies of $\mathrm{SL}(2, \mathbb{Z})$ contains as a subgroup of finite index the direct product D of n non-abelian free groups. D embeds in the mapping class group of a surface of genus n with 1 boundary component, so Γ embeds in Mod(S) for some closed surface S. It is proved in [20] that Γ contains finitely presented subgroups with infinitely many conjugacy classes of elements of order 4. (See also [15].) ∎

3.3 Fibre products and subdirect products of free groups

We shall develop our discussion of these subgroups around two constructions.

Construction 3.7 (Fibre products). *Let $Q = \langle A \mid R \rangle$ be a finitely presented group. Let F be the free group on A and let $p : F \to Q$ be the surjection implicit in the notation. The* *kernel of p is finitely generated if and only if Q is finite, but, regardless of what Q is, the fibre product*

$$P = \{(u, v) \mid p(u) = p(v)\} < F \times F$$

will be generated by the finite set $\Sigma_Q = \{(a, a), (r, 1); a \in A, r \in R\}$.

This observation provides complicated finitely generated subgroups of $F \times F$: Mihailova [57] and Miller [55] were the first to see that if Q has an unsolvable word problem, then P has an unsolvable conjugacy problem (cf. Proposition 3.37) and there is no algorithm to decide which words in the generators of $F \times F$ define elements of P (Proposition 3.36); and since $P \cong F \times F$ if and only if $Q = 1$, there can be no algorithm to decide isomorphism among the finitely generated subgroups of $F \times F$, because there is no algorithm that can determine which finitely presented groups (with generating sets of a fixed cardinality) are trivial. There isn't even an algorithm that, given a finite subset $\Sigma \subset F \times F$, can calculate the first homology of $\langle \Sigma \rangle$ (see [27]).

There are uncountably many 2-generator groups, and via fibre products one can deduce from this that there are uncountably many non-isomorphic subgroups $P < F \times F$ (see [10]). Hence, we have the following.

Proposition 3.8. *The mapping class group of any surface of genus at least 2 contains uncountably many non-isomorphic subgroups.*

These fibre products, though, do not give us complicated *finitely presented* subgroups of mapping class groups, because P is finitely presentable if and only if Q is finite [45]. Indeed, Baumslag and Roseblade [10] proved that $F \times F$ has no finitely presented subgroups other than the obvious ones: if $G < F \times F$ is finitely presented, then either G is free or else it has a subgroup of finite index that is a product of two free groups (its intersections with the direct factors).

3.3.1 Finitely presented examples

A celebrated construction of Stallings [65] and Bieri [12] shows that there are interesting finitely presented subgroups in the direct product of three (or more) free groups.

Construction 3.9 (The Stallings–Bieri groups). *Let $h : F \times \cdots \times F \to \mathbb{Z}$ be a homomorphism that restricts to an epimorphism on each of the n factors. Stallings [65] (in the case n = 3) and Bieri [12] proved that the kernel SB_n has a classifying space with a finite $(n - 1)$-skeleton, but $H_n(SB_n, \mathbb{Z})$ is not finitely generated.*

In the light of the discussion following Construction 3.7, one might anticipate that these examples are the tip of an iceberg of pathology akin to the wildness that we saw among the finitely generated subgroups of $F \times F$. But Bridson and Miller [28] (cf. [25]) proved, roughly speaking, that variations on the construction of Stallings and Bieri account for all the finitely presented subdirect products of free groups. Recall that a subgroup of a direct product $H < G_1 \times \cdots \times G_d$ is termed a *subdirect product* if its projection to each of the factors is onto, and H is said to be *full* if all of the intersections $H \cap G_i$ are non-trivial.

Theorem 3.10 (from [28]). *If $H < F_1 \times \cdots \times F_n = D$ is a full subdirect product of finitely generated free groups, then there is a subgroup of finite index $D_0 < D$ such that H contains the $(n-1)$st term of the lower central series of D_0.*

3.4 A new level of complication

We now come to the cluster of new results concerning subgroups of mapping class groups. These theorems show that the finitely presented subgroups of mapping class groups can be vastly more complicated than those studied hitherto. The proofs of these theorems will be outlined in later sections, where they are used to illustrate a general technique for constructing wild subgroups in varied contexts.

Readers unfamiliar with decision problems may wish to consult [56]. For an account of the history of the problems settled by the following theorems, see [40].

Theorem 3.11 (from [16]). *If the genus of S is sufficiently large, then the isomorphism problem for the finitely presented subgroups of $Mod(S)$ is unsolvable.*

In more detail, there is a recursive sequence Δ_i $(i \in \mathbb{N})$ of finite subsets of $Mod(S)$, together with finite presentations $\langle \Delta_i \mid \Theta_i \rangle$ of the subgroups they generate, such that there is no algorithm that can determine whether or not $\langle \Delta_i \mid \Theta_i \rangle \cong \langle \Delta_0 \mid \Theta_0 \rangle$.

Theorem 3.12 (from [16]). *If the genus of S is sufficiently large, then there is a finitely presented subgroup of $Mod(S)$ with unsolvable conjugacy problem.*

Theorem 3.13 (from [16]). *If the genus of S is sufficiently large, then there are finitely presented subgroups of $Mod(S)$ for which the membership problem is unsolvable.*

The *Dehn function* of a finitely presented group $\Gamma = \langle A \mid R \rangle$ estimates the complexity of the word problem by counting the number of times one has to apply the defining relations in order to prove that a word w in the generators represents the identity in the group: $\text{Area}(w)$ is defined to be the least integer N for which there is an equality

$$w = \prod_{i=1}^{N} \theta_i r_i^{\pm 1} \theta_i^{-1}$$

in the free group $F(A)$, with $r_i \in R$, and the Dehn function of $\langle A \mid R \rangle$ is

$$\delta(n) := \max\{\text{Area}(w) \mid w =_\Gamma 1, |w| \le n\},$$

where $|w|$ denotes word length. Mapping class groups have quadratic Dehn functions [60], as do the finitely presented subgroups described in the previous sections.

Theorem 3.14 (from [16]). *If the genus of S is sufficiently large, then there are finitely presented subgroups of $Mod(S)$ whose Dehn functions are exponential.*

One might hope to prove some of these theorems by focusing on subgroups of direct products of free groups, but restrictions that follow from Theorem 3.10 dash this hope. The following theorem was proved in [28] and extended to all finitely presented subgroups of residually free groups in [26].

Theorem 3.15 (from [28]). *The conjugacy problem is solvable for every finitely presented subgroup of a direct product of free (or surface) groups, and there is a uniform solution to the membership problem for all such subgroups.*

In Section 3.6, we consider how we might enlarge the class of direct products of free groups so as to obtain wilder subgroups (via fibre product constructions) while retaining enough geometry to provide embeddings into mapping class groups. But first we turn to a different topic.

3.5 The nilpotent genus of a group

This section is based on joint work with Alan Reid from the University of Texas [29].

If each finite subset of a group Γ injects into some nilpotent (or finite) quotient of Γ, then one expects to be able to detect many properties of Γ from the totality of its nilpotent (or finite) quotients. Which properties can be detected and which cannot? Forms of this question have stimulated a lot of research into discrete and profinite groups over the last forty years, and there has been a particular resurgence of interest recently, marked by several notable breakthroughs. Here we focus on the nilpotent quotients.

Recall that a group Γ is said to be *residually nilpotent* (respectively *residually torsion-free nilpotent*) if for each non-trivial $\gamma \in \Gamma$ there exists a nilpotent (respectively torsion-free nilpotent) group Q and a homomorphism $\phi : \Gamma \to Q$ with $\phi(\gamma) \neq 1$. Thus Γ is residually nilpotent if and only if $\bigcap \Gamma_n = 1$, where Γ_n is the nth term of the *lower central series* of Γ, defined inductively by setting $\Gamma_1 := \Gamma$ and $\Gamma_{n+1} := \langle [x, y] : x \in \Gamma_n, y \in \Gamma \rangle$.

We say that two residually nilpotent groups Γ and Λ have the same *nilpotent genus* if they have the *same nilpotent quotients*; this is equivalent to requiring that $\Gamma/\Gamma_c \cong \Lambda/\Lambda_c$ for all $c \geq 1$.

Example 3.16. *Examples of finitely generated residually torsion-free nilpotent groups include free groups F_n (and hence residually free groups such as surface groups and limit groups), right-angled Artin groups (RAAGs) [38], the Torelli subgroup of the mapping class group [5] and $IA_n < \mathrm{Out}(F_n)$, the kernel of the natural map $\mathrm{Out}(F_n) \to \mathrm{GL}(n, \mathbb{Z})$ (see [3, 5, 30]), and the corresponding subgroup in the outer automorphism group of any RAAG [66].*

A group is termed parafree *if it is residually nilpotent and has the same genus as a free group. The existence of families of parafree groups that are not free gives a first inkling of the diversity that can exist within a fixed nilpotent genus. One such family was discovered by Gilbert Baumslag (see [6]):*

$$G_{ij} = \langle a, b, c \mid a = [c^i, a].[c^j, b] \rangle.$$

In [7], Baumslag surveyed the state of the art concerning groups of the same nilpotent genus and compiled a list of open problems that are of particular importance in the field. Here we shall concentrate on three problems whose resolution will serve to emphasize how different groups within a given genus can be. (Other problems on Baumslag's list challenge the reader to establish commonalities across a genus; cf. [29, Theorem C].)

Problems 3.17. *Do there exist pairs of groups of the same nilpotent genus such that*

- *one is finitely presented and the other is not; or*
- *both are finitely presented, one has a solvable conjugacy problem but the other does not; or*
- *one has finitely generated second homology $H_2(-, \mathbb{Z})$ and the other does not?*

These questions are settled by the following compilation of results from [29].

Theorem 3.18.

1. *There exist pairs of finitely presented, residually torsion-free nilpotent groups $P \hookrightarrow \Gamma$ of the same nilpotent genus such that Γ has a solvable conjugacy problem and P does not.*

2. *There exist pairs of finitely generated, residually torsion-free nilpotent groups $N \hookrightarrow \Gamma$ of the same nilpotent genus such that Γ is finitely presented while $H_2(N, \mathbb{Q})$ is infinite-dimensional (so, in particular, N is not finitely presented).*

Remark 3.19. The pairs of groups that are constructed in [29] to prove this theorem have the additional property that the inclusion map induces an isomorphism of profinite and pro-nilpotent completions.

3.5.1 Criteria for pro-nilpotent equivalence

The following theorem of John Stallings [64] provides a useful criterion for establishing that groups have the same nilpotent genus: *If a homomorphism of groups $u : N \to \Gamma$ induces an isomorphism on $H_1(-, \mathbb{Z})$ and an epimorphism on $H_2(-, \mathbb{Z})$, then $u_c : N/N_c \to \Gamma/\Gamma_c$ is an isomorphism for all $c \geq 1$.*

Given a short exact sequence of groups $1 \to N \to G \to Q \to 1$, the Lyndon–Hochschild–Serre (LHS) spectral sequence (which is explained in [31, p. 171]) calculates the homology of G in terms of the homology N and Q. The terms on the E^2 page of the spectral sequence are $E^2_{pq} = H_p(Q, H_q(N, \mathbb{Z}))$, where the action of Q on $H_*(N, \mathbb{Z})$ is induced by the action of G on N by conjugation. The following proposition is proved in [29] by using this spectral sequence to see that $N \to \Gamma$ satisfies the hypotheses of Stallings' theorem.

Proposition 3.20. *Let $1 \to N \overset{u}{\to} \Gamma \to Q \to 1$ be a short exact sequence of groups and let $u_c : N/N_c \to \Gamma/\Gamma_c$ be the homomorphism induced by $u : N \hookrightarrow \Gamma$. Suppose that N is finitely generated, that Q has no non-trivial finite quotients and that $H_2(Q, \mathbb{Z}) = 0$. Then*

u_c is an isomorphism for all $c \geq 1$. In particular, if Γ is residually nilpotent, then N and Γ have the same nilpotent genus.

Corollary 3.21. *Under the hypotheses of Proposition 3.20, the inclusion $P \hookrightarrow \Gamma \times \Gamma$ of the fibre product induces an isomorphism $P/P_c \to \Gamma/\Gamma_c \times \Gamma/\Gamma_c$ for every $c \in \mathbb{N}$ and hence an isomorphism $\widehat{P}_{nil} \to \widehat{\Gamma}_{nil} \times \widehat{\Gamma}_{nil}$.*

Proof. We have inclusions $N \times N \xrightarrow{i} P \xrightarrow{j} \Gamma \times \Gamma$ and the proposition implies that i and $j \circ i$ induce isomorphisms modulo any term of the lower central series, and therefore j does as well. ∎

3.5.2 Relation with profinite genus

Let Γ be a finitely generated group. If one orders the normal subgroups of finite index $N < \Gamma$ by reverse inclusion, then the quotients Γ/N form an inverse system whose limit

$$\widehat{\Gamma} = \varprojlim \Gamma/N$$

is the *profinite completion* of Γ. Similarly, the *pro-nilpotent completion*, denoted by $\widehat{\Gamma}_{nil}$, is the inverse limit of the nilpotent quotients of Γ. Every homomorphism of discrete groups $u : H \to G$ induces a homomorphism $\hat{u} : \widehat{H} \to \widehat{\Gamma}$ and a homomorphism $\hat{u}_{nil} : \widehat{H}_{nil} \to \widehat{\Gamma}_{nil}$.

If H, G are residually nilpotent, they lie in the same nilpotent genus if and only if $\widehat{H}_{nil} \cong \widehat{G}_{nil}$. If G is nilpotent, then $G = \widehat{G}_{nil}$.

There are finitely generated nilpotent groups $H \not\cong G$ that have the same finite quotients; thus $\widehat{H} = \widehat{G}$ but $\widehat{H}_{nil} \not\cong \widehat{G}_{nil}$. The situation is quite different if the isomorphism of profinite completions is induced by a homomorphism of the discrete groups.

Proposition 3.22 (from [29]). *Let $u : P \hookrightarrow \Gamma$ be a pair of finitely generated, residually finite groups, and for each $c \geq 1$, let $u_c : P/P_c \to \Gamma/\Gamma_c$ be the induced homomorphism. If $\hat{u} : \widehat{P} \to \widehat{\Gamma}$ is an isomorphism, then u_c is an isomorphism for all $c \geq 1$, and hence $\widehat{P}_{nil} \cong \widehat{\Gamma}_{nil}$.*

With this observation in hand, we see that (modulo variations in the finiteness assumptions) Proposition 3.20 is a weak form of the following proposition, which originates in the work of Platonov and Tavgen [61], where the first pairs of finitely generated groups $P \hookrightarrow \Gamma$ satisfying the hypotheses of Proposition 3.22 were constructed. (Such pairs of groups are now known as *Grothendieck pairs*.) This proposition also played an important role in the Bridson–Grunewald construction of Grothendieck pairs where both P and Γ are finitely presented [23].

Proposition 3.23. *Let $1 \to N \to \Gamma \to Q \to 1$ be a short exact sequence of groups with Γ finitely generated and let P be the associated fibre product. Suppose that $Q \neq 1$ is finitely presented and has no proper subgroups of finite index, and that $H_2(Q, \mathbb{Z}) = 0$. Then*

1. *$P \to \Gamma \times \Gamma$ induces an isomorphism $\widehat{P} \to \widehat{\Gamma \times \Gamma}$;*
2. *if N is finitely generated, then $N \to \Gamma$ induces an isomorphism $\widehat{N} \to \widehat{\Gamma}$.*

3.6 Cubes, RAAGs and CAT(0)

In this section, we shall see *right-angled Artin groups* (RAAGs) emerge as a general-ization of direct products of free groups. RAAGs have a similar cubical geometry to $F \times \cdots \times F$ and are residually torsion-free nilpotent; moreover, it is easy to get them to act on surfaces. But, crucially for us, they harbour a much greater array of finitely presented subgroups.

We take up the theme of Construction 3.9, retaining the notation.

The original proofs of Stallings and Bieri are essentially algebraic. Bestvina and Brady [11] discovered a geometric proof that motivated their Morse theory for cubical com-plexes. If we regard F as the fundamental group of a compact simplicial graph Y, then $D = F \times \cdots \times F$ is the fundamental group of $X = Y \times \cdots \times Y$, which has a natural cubical structure. This **cube complex** is non-positively curved in the sense of Alexandrov, i.e. locally CAT(0). (The standard reference for CAT(0) spaces is [24], but [32] and [63] cover more than enough for our needs.)

The vertex set of the universal cover is D and the homomorphism $h : D \to \mathbb{Z}$ can be extended linearly across cells to give a *Morse function* $\tilde{X} \to \mathbb{R}$. Bestvina and Brady [11] determined the finiteness properties of the kernel of h by examining the way in which the sublevel sets of this Morse function change as one passes through critical points (ver-tices). They extended this analysis to the larger class of cubical complexes defined below, and in this way settled long-standing questions concerning the relationship between different finiteness properties of groups.

A *right-angled Artin group* (RAAG) is a group given by a presentation of the form

$$A = \langle v_1, \dots, v_n \mid [v_i, v_j] = 1 \ \forall (i, j) \in E \rangle.$$

Thus A is encoded by a graph with vertex set $\{v_1, \dots, v_n\}$ and edge set $E \subset V \times V$. The prototype $F_2 \times \cdots \times F_2$ is the RAAG associated to the 1-skeleton of the simplicial join $\mathbb{S}^0 * \cdots * \mathbb{S}^0$. The *Salvetti complex* is the classifying space for A obtained by gluing stand-ard tori (cubes with opposite faces identified) along coordinate faces according to the commuting relations in the presentation; it has non-positive curvature.

I have portrayed RAAGs as a natural generalization of direct products of free groups, but in many ways this fails to do them justice. They have gained prominence in recent years as an extremely important class of groups whose simple description belies their rich structure. From the point of view of this chapter, their three most important features are the richness of their subgroup structure, the ease with which they can be made to act on a great range of objects and their residual properties.

3.6.1 RAAGs everywhere

Whenever one has n automorphisms α_i of an object X, some of which commute, say $[\alpha_i, \alpha_j] = 1$ if $(i, j) \in E$, then one has an action of the RAAG associated to the n-vertex graph with edge-set E. Roughly speaking, this action will be faithful if the α_i that do not commute are unrelated. One such setting is that of surface automorphisms: if two simple closed curves on a surface are disjoint, then the Dehn twists in those curves commute, but if one has a set of curves no pair of which can be homotoped off each other, then suitable powers of the twists in those curves freely generate a free group. (Significantly sharper results of this sort are proved in [52] and [33].) It follows that any RAAG can be embedded in the mapping class group of any surface S of sufficiently high genus: it suffices that the dual of the graph defining A can be embedded in S. (This is explained by Crisp and Wiest in [35].) The surface can have boundary and punctures. With more care, one can arrange for the embedding of the RAAG to lie in the Torelli subgroup $\mathcal{T}(S) <$ Mod(S) (cf. [52]).

Proposition 3.24. *Every RAAG embeds in the mapping class group of any surface of sufficiently high genus.*

3.6.2 Some properties of RAAGs

With the preceding proposition in hand, we see that the theorems stated in Section 3.4 will follow if we can construct RAAGs with subgroups of the desired kind. Similarly, we shall solve Baumslag's problems (3.17) by constructing suitable RAAGs and exploiting the following theorem proved by Droms in his thesis [38] (cf. [39, 67]).

Theorem 3.25 (from [38]). *RAAGs are residually torsion-free nilpotent.*

Further important properties of RAAGs include the fact that they are linear [49] even over $\mathbb{Z}$ [37], they are conjugacy separable [58], they are residually finite rational solvable (RFRS) in the sense of Agol [2] and their quasiconvex subgroups are virtual retracts (and so are closed in the profinite topology) [46].

3.6.3 Special cube complexes

What has really brought RAAGs to the fore in recent years is the richness of their subgroup structure. We saw hints of this in the work of Bestvina and Brady, but the spectacular extent of this richness truly emerged from the theory of *special cube complexes* initiated by Haglund and Wise [47] and advanced in many subsequent papers, particularly by Wise and his coauthors. (There is much more on this in [63].)

Definition 3.26. *A non-positively curved cube complex X is **special** if it admits a locally isometric embedding into the Salvetti complex $K(A, 1)$ of a RAAG A.*

Remark 3.27. A locally isometric map between compact non-positively curved spaces induces an injective map on fundamental groups [24, p. 201], so the fundamental groups of special cube complexes are subgroups of RAAGs.

The definition of *special* is not a very practical one, but Haglund and Wise [47] prove that it is equivalent to a short list of conditions on the behaviour of hyperplanes in the given cube complex (see [63]).

Theorem 3.28 (from [47]). *A non-positively curved cube complex is special if and only if its hyperplanes are 2-sided, do not self-cross, do not self-osculate and do not inter-osculate.*

This remarkable insight makes it possible to verify specialness (of X or some finite cover of it) in many instances. Thus we have a putative machine for constructing interesting subgroups of RAAGs (hence residually torsion-free nilpotent groups that are subgroups of mapping class groups):

- **Cubulate** groups, i.e. find methods for exhibiting large classes of groups as fundamental groups of compact non-positively curved cube complexes. (This is the central theme of [63].)

- Use the Haglund–Wise criterion to prove that these cube complexes X are special, or at least that finite-sheeted covers of them are special (i.e. X is **virtually special**).

This programme, widely promoted by Dani Wise, has proved extremely successful. Two results are of particular importance for our purposes, one for each of the steps articulated above. To state the first, we need the vocabulary of small-cancellation groups.

The symmetrization R^* of a set of words R over an alphabet X consists of all cyclic permutations of w and w^{-1} with $w \in R$. The set R and the presentation $\langle X \mid R \rangle$ are said to satisfy the $C'(\lambda)$ *small-cancellation condition* if pieces (i.e. recurring subwords) are bounded in length: if there exist distinct $w, w' \in R^*$ such that $w \equiv uv$ and $w' \equiv uv'$, then $|u| < \lambda|w|$. If R is finite and $\lambda \leq \frac{1}{6}$, then $\Gamma \cong \langle X \mid R \rangle$ is hyperbolic (in the sense of Gromov) and torsion-free.[3]

Theorem 3.29 (from [68]). *If a group G has a finite presentation that is $C'(\frac{1}{6})$, then G acts properly and cocompactly on a $CAT(0)$ cube complex.*

Many groups are now proved to be **virtually special**. The crowning achievement, following much work of Wise [69] and others, is Agol's theorem.

Theorem 3.30 (from [1]). *Let X be a compact non-positively curved cube complex. If $\pi_1 X$ is hyperbolic, then X is virtually special.*

This theorem has many consequences—the most important to date being Agol's resolution of the virtual fibering conjecture for hyperbolic 3-manifolds (cf. Section 3.11)—but the two that concern us here are the following.

Corollary 3.31. *If a hyperbolic group H is the fundamental group of a compact non-positively curved cube complex, then H embeds in the mapping class group of infinitely many (closed) surfaces, and some subgroup of finite-index $H_0 < H$ is residually torsion-free nilpotent.*

Proof. The first assertion follows from Proposition 3.24 and Corollary 3.4, the second from Theorem 3.25. ∎

3.6.4 The mapping class genus of virtually special groups

Corollary 3.31 assures us that the following quantity, which one might call the *mapping class genus*, is a well-defined invariant of hyperbolic groups that can be cubulated. It seems difficult to compute, but may well provide a rich field of exploration. The case of Kleinian groups is already interesting. Here S_g denotes the closed orientable surface of genus g:

$$\mathrm{mcg}(\Gamma) := \min\{g \mid \exists\, \Gamma \hookrightarrow \mathrm{Mod}(S_g)\}.$$

3.7 Rips, fibre products and 1-2-3

We take up the theme of Construction 3.7. The lack of finite presentability in the fibre products that we constructed $P < F \times F$ can be traced to the fact that a non-trivial normal subgroup of infinite index in a free group cannot be finitely generated. The key to getting around this problem is to express groups as quotients of hyperbolic groups instead of free groups, with a gain in the finiteness properties of the kernel. This idea is due to E. Rips [62].

Theorem 3.32 (from [62]). *There is an algorithm that, given a finite group-presentation $\mathcal{Q}$, will construct a short exact sequence*

$$1 \to N \to H \xrightarrow{q} Q \to 1$$

where H is $C'(\frac{1}{6})$ small-cancellation, Q is the group with presentation $\mathcal{Q}$, and N is a 2-generator group.

Remark 3.33. The discussion in Section 3.6 shows that H is virtually special. In fact, one can arrange this more directly: the Rips construction is very flexible and variations by different authors have imposed extra conditions on H; before Agol's work, Haglund and Wise [47] used this flexibility to arrange for H to be virtually special (at the cost of adding more generators to N).

In the spirit of Construction 3.7, we focus on the fibre product $P = \{(h, h') \in H \times H \mid q(h) = q(h')\}$. We want P to be finitely presented, but in general it will not be (cf. Lemma 3.38). However, if Q is of *type F_3* (i.e. has a classifying space $K(Q, 1)$ with finite 3-skeleton), then the **1-2-3 Theorem** of [8] assures us that P *will* be finitely presented.

Theorem 3.34 (from [8]). *Let $1 \to N \to \Gamma \xrightarrow{q} Q \to 1$ be a short exact sequence of groups. If N is finitely generated, Γ is finitely presented and Q is of type F_3, then the associated fibre product $P < \Gamma \times \Gamma$ is finitely presented.*

One may wonder if N can also be made finitely presented—but it cannot.

Lemma 3.35. *If Q is infinite, then the subgroup $N < \Gamma$ in the Rips construction is not finitely presented.*

Proof. Being a small-cancellation group, Γ has cohomological dimension 2. Bieri [13] proved that a finitely presented normal subgroup of infinite index in a group of cohomological dimension 2 must be free. But in the Rips construction, N is visibly not free: it is a 2-generator group that has non-trivial relations, and it is not cyclic because non-elementary hyperbolic groups do not have cyclic normal subgroups. ∎

3.7.1 Finite and nilpotent quotients

Proposition 3.20 and its corollary show that if Q has no finite quotients and $H_2(Q, \mathbb{Z}) = 0$, then the pair of groups $N \xrightarrow{u} \Gamma$ produced by the Rips construction is such that u induces an isomorphism modulo each term of the lower central series, and so does the inclusion of the fibre product $P \hookrightarrow \Gamma \times \Gamma$. Likewise, Proposition 3.23 tells us that these inclusions induce isomorphisms of profinite completions.

3.7.2 Rips translates foibles from Q to N and P

The translation of properties from Q to N and P has to be analysed according to context. The ones that interest us here concern finite and nilpotent quotients (as described above), decision problems and finiteness conditions.

   If Q has an unsolvable word problem, this manifests itself in unsolvable decision problems of a different type for N and P.

Proposition 3.36 (from [8]). *Let* $1 \to N \to \Gamma \xrightarrow{p} Q \to 1$ *be a short exact sequence of groups, with* Γ *finitely generated, and let* $P < \Gamma \times \Gamma$ *be the associated fibre product. If the word problem in* Q *is unsolvable, then the membership problem for* $P < \Gamma \times \Gamma$ *is unsolvable.*

Proof. We fix a finite generating set X for Γ and work with the generators $X' = \{(x, 1), (1, x) \mid x \in X\}$ for $\Gamma \times \Gamma$. Given a word $w = x_1 \dots x_n$ in the free group on X, we consider the word $(x_1, 1) \dots (x_n, 1)$ in the free group on X'. This word defines an element of P if and only if $p(w) = 1$ in Q, and we are assuming that there is no algorithm that can determine which words in the symbols $p(x)$ equal the identity in Q. ∎

Proposition 3.37 (from [8]). *Let* $1 \to N \to \Gamma \xrightarrow{p} Q \to 1$ *be a short exact sequence of groups, with* Γ *torsion-free and hyperbolic, and let* $P < \Gamma \times \Gamma$ *be the associated fibre product. If the word problem in* Q *is unsolvable, then the conjugacy problem is unsolvable in* N *and in* P.

Sketch of proof.   Fix finite generating sets B for Γ and A for N. Fix $a \in N \smallsetminus \{1\}$. For each $b \in B$ and $\epsilon = \pm 1$, let $u_{b,\epsilon}$ be a word in the free group on A so that $b^\epsilon a b^{-\epsilon} = u_{b,\epsilon}$ in G. Given an arbitrary word w in the letters B, one can use the relations $b^\epsilon a b^{-\epsilon} = u_{b,\epsilon}$ to convert waw^{-1} into a word w' in the letters A. Now ask if w' is conjugate to a in N. The answer is 'yes' if $w \in N$, and consideration of centralizers shows that it is 'no' if $w \notin N$. Thus w' is conjugate to a in N if and only if $p(w) = 1$ in Q, and we are assuming that there is no algorithm that can decide if this is the case.
The argument for P is similar but more involved; see [8, Section 3]. ∎

The following lemma is proved using the LHS spectral sequence; see [29, Section 6].

Lemma 3.38. *Let* $1 \to N \to G \to Q \to 1$ *be a short exact sequence of finitely gener-ated groups. If* $H_3(G, \mathbb{Q})$ *is finite dimensional but* $H_3(Q, \mathbb{Q})$ *is infinite dimensional, then* $H_2(N, \mathbb{Q})$ *is infinite-dimensional.*

3.8 Examples template

The following template can be employed in any context where one is interested in demonstrating diverse or extreme behaviour among the finitely presented subgroups of groups in a class of groups C with the property that every RAAG embeds in some $\Gamma \in C$.

- Feed *designer groups* Q into the *Rips construction* to obtain

$$1 \to N \to H \to Q \to 1.$$

- Pass to a subgroup of finite index in H to obtain

$$1 \to N_0 \to H_0 \xrightarrow{q} Q_0 \to 1,$$

  N_0 finitely generated, H_0 special (*subgroup of a RAAG*) and $Q_0 < Q$ finite-index.
- Pass to the *fibre product*

$$P = \{(h, h') \in H_0 \times H_0 \mid q(h) = q(h')\}$$

  and note that P is a subgroup of a RAAG.
- Note that by the *1-2-3 Theorem*, P is finitely presented[4] if Q is of type F_3.
- Embed the RAAG containing P in $\Gamma \in C$.

This is a rather general and loosely stated template, but it does have remarkably wide applicability. We shall use it to resolve the problems for mapping class groups and nilpotent genus that are the focus of our story. But to make use of the template, we have to resolve the following difficulties:

- If we are interested in constructing finitely presented groups $P < H_0 \times H_0$ with some property $\mathcal{P}$, then we must first identify a related property $\overline{\mathcal{P}}$ such that if Q has $\overline{\mathcal{P}}$, then $P < H_0 \times H_0$ has $\mathcal{P}$ (cf. Section 3.7.2).
- We have to prove that there exist groups of type F_3 with property $\overline{\mathcal{P}}$.
- We have to ensure that $\overline{\mathcal{P}}$ is inherited by subgroups of finite index $Q_0 < Q$; this might be inherent to $\overline{\mathcal{P}}$, but we might have to control the finite-index subgroups of Q, perhaps arranging that there are none other than Q itself; cf. [21, 23].

3.8.1 Crafting designer groups

The construction of input groups Q for the template is very particular to the situation at hand and typically requires ad hoc innovation. In order to prove Theorems 3.12, 3.13 and 3.18, we need groups with the following properties:

Examples 3.39.

1. *There are infinite groups of type F_3 that have no non-trivial finite quotients. The first such examples were constructed by Graham Higman [48], and a general method for constructing such groups is described in [21] and [23].*

2. *There are groups of type F_3 that have an unsolvable word problem. Examples of this sort were constructed by Collins and Miller [34].*

3. *There exist infinite groups Q of type F_3 that, simultaneously, have no non-trivial finite quotients and an unsolvable word problem and for which $H_2(Q, \mathbb{Z}) = 0$; see [22, Theorem 3.1].*

4. *There are finitely presented groups Δ with no non-trivial finite quotients, so that $H_i(\Delta, \mathbb{Z}) = 0$ for $i = 1, 2$ but $\dim H_3(\Delta, \mathbb{Q}) = \infty$. Such groups are constructed in [29].*

3.9 Proofs from the template

Proposition 3.24 and Corollary 3.4 tell us that Theorems 3.13 and 3.12 will follow if we can prove the same results for subgroups of RAAGs or virtually special groups. Likewise, since RAAGs are residually torsion-free nilpotent, in order to prove Theorem 3.18, it is enough to exhibit RAAGs with the stated properties. In each case, we shall use the examples template from Section 3.8 to construct suitable RAAGs. I shall state the key points and I encourage the reader to check the details.

3.9.1 Proof of Theorem 3.13 and 3.12

Let property $\overline{\mathcal{P}}$ be the insolubility of the word problem. Apply the template with Q as in Example 3.39(2) and appeal to Propositions 3.36 and 3.37.

3.9.2 Proof of Theorem 3.18(1)

Let property $\overline{\mathcal{P}}$ be the insolubility of the word problem. Apply the template with Q as in Example 3.39(3) and appeal to Corollary 3.21 and Proposition 3.37.

3.9.3 Proof of Theorem 3.18(2)

Apply the template with Q as in Example 3.39(4) and appeal to Proposition 3.20 and Lemma 3.38.

3.10 The isomorphism problem for subgroups of RAAGs and Mod(*S*)

To prove Theorem 3.11, we use a criterion due to Bridson and Miller [27].

Theorem 3.40 (from [27]). *Let* $1 \to N \to \Gamma \to L \to 1$ *be an exact sequence of groups. Suppose that*

1. Γ *is torsion-free and hyperbolic,*

2. N *is infinite and finitely generated, and*

3. L *is a non-abelian free group.*

If F *is a non-abelian free group, then the isomorphism problem for finitely presented subgroups of* $\Gamma \times \Gamma \times F$ *is unsolvable.*

In more detail, there is a recursive sequence Δ_i $(i \in \mathbb{N})$ *of finite subsets of* $\Gamma \times \Gamma \times F$, *together with finite presentations* $\langle \Delta_i \mid \Theta_i \rangle$ *of the subgroups they generate, such that there is no algorithm that can determine whether or not* $\langle \Delta_i \mid \Theta_i \rangle \cong \langle \Delta_0 \mid \Theta_0 \rangle$.

The subgroup G_i presented by $\langle \Delta_i \mid \Theta_i \rangle$ is obtained by fixing a splitting $\Gamma = N \rtimes L$ and defining G_i to be the subgroup generated by $N \times N$ and $\{(\phi_i(x), x) \mid x \in F\}$, where $\phi_i : F \to L \times L$ is a homomorphism whose image is a subdirect product. A key feature of the construction in $[27]$ is that the finiteness properties of centralizers in G_i are intimately connected to the question of whether ϕ_i is onto. We saw in Section 3.3 that there is no algorithm that can determine if a finite subset of a direct product of free groups generates the product, and this provides a seed of undecidability that propagates through the construction.

Corollary 3.41. *There exist RAAGs in which the isomorphism problem for finitely presented subgroups is unsolvable.*

Proof. We apply the template of Section 3.8 with Q a non-abelian free group and define $\Gamma = H_0$. If H_0 is a subgroup of the RAAG A, then $\Gamma \times \Gamma \times F$ will be a subgroup of the RAAG $A \times A \times F$. ∎

In the light of Proposition 3.24, Theorem 3.11 follows from this corollary.

Remark 3.42. Baumslag and Miller $[9]$ proved that the isomorphism problem is unsolvable in the class of finitely presented residually torsion-free nilpotent groups. Corollary 3.41 provides an alternative proof.

3.11 Dehn functions

One can prove Theorem 3.14 without the Rips construction or the 1-2-3 Theorem: one can deduce it directly from the fact that the fundamental groups of closed hyperbolic 3-manifolds are virtually special, applying Proposition 3.24 to embed the RAAG A of the following proposition into mapping class groups.

Proposition 3.43. *There exist RAAGs* A *and finitely presented subgroups* $P < A$ *such that* P *has an exponential Dehn function.*

Proof. Let M be a closed, orientable, hyperbolic 3-manifold that fibres over the circle. Then $\pi_1 M = \Sigma \rtimes \mathbb{Z}$, where Σ is the fundamental group of a closed surface of genus at least 2, and $\Gamma = \pi_1 M \times \pi_1 M$ contains $P := (\Sigma \times \Sigma) \rtimes \mathbb{Z}$, the inverse image of the diagonal in $\Gamma/(\Sigma \times \Sigma) = \mathbb{Z} \times \mathbb{Z}$. The Dehn function of P is exponential; see [19, Theorem 2.5]. The growth of a Dehn function is preserved on passage to subgroups of finite index, and by [1] there is a subgroup of finite index in Γ that embeds in a RAAG. ∎

Acknowledgements

The author was a Clay Senior Scholar in Residence at the Park City Mathematics Institute in 2012. He thanks the Clay Mathematics Institute for this honour. He also thanks the Royal Society for the Wolfson Research Merit Award that supports his research.

Notes

1. This reflects the more historic name, Teichmüller modular group.
2. Most of the results of Ivanov that I quote were proved in his earlier papers, but [50] provides an excellent, coherent account of his work.
3. In saying this, I am skipping over a technicality about how to treat words in R that are proper powers.
4. If there is an algorithm to construct a finite 3-skeleton for a $K(Q, 1)$, then one can construct a finite presentation for P in an algorithmic manner, but this requires further argument [26].

References

[1] I. Agol, *The Virtual Haken Conjecture* (with appendix by I. Agol, D. Groves and J. Manning), Doc. Math. **18** (2013) 1045–1087. MR3104553.

[2] I. Agol, *Criteria for virtual fibering*, J. Topol. **1** (2008) 269–284. MR2399130 (2009b:57033).

[3] S. Andreadakis, *On the automorphisms of free groups and free nilpotent groups*, Proc. Lond. Math. Soc. **15** (1965) 239–268. MR0188307 (32:5746).

[4] M. Aschenbrenner, S. Friedl and H. Wilton, *3-Manifold Groups*, EMS Series of Lectures in Mathematics, European Mathematical Society (EMS), Zürich, 2015. MR3444187.

[5] H. Bass and A. Lubotzky, *Linear-central filtrations on groups*, in W. Abikoff, J. S. Birman and K. Kuiken, eds., *The Mathematical Legacy of Wilhelm Magnus: Groups, Geometry and Special Functions* (American Mathematical Society, 1994), pp. 45–98. MR1292897 (96c:20054).

[6] G. Baumslag, *Musings on Magnus*, in W. Abikoff, J. S. Birman and K. Kuiken, eds., *The Mathematical Legacy of Wilhelm Magnus: Groups, Geometry and Special Functions* (American Mathematical Society, 1994), pp. 99–106. MR1292898 (95i:20038).

[7] G. Baumslag, *Parafree groups*, in L. Bartholdi, T. Ceccherini-Silberstein, T. Smirnova-Nagnibeda and A. Zuk, eds. *Infinite Groups: Geometric, Combinatorial and Dynamical Aspects* (Birkhäuser, 2005), pp. 1–14. MR2195450 (2006j:20039).

[8] G. Baumslag, M. R. Bridson, C. F. Miller III and H. Short, *Fibre products, non-positive curvature, and decision problems*, Comment. Math. Helv. **75** (2000) 457-477. MR1793798 (2001k:20091).

[9] G. Baumslag and C.F. Miller III, *The isomorphism problem for residually torsion-free nilpotent groups*, Groups Geom. Dyn. **1** (2007) 1-20. MR2294245 (2008f:20062).

[10] G. Baumslag and J. Roseblade, *Subgroups of direct products of free groups*, J. Lond. Math. Soc. (2) **30** (1984) 44-52. MR760871 (86d:20028).

[11] M. Bestvina and N. Brady, *Morse theory and finiteness properties of groups*, Invent. Math. 129 (1997) 445-470. MR1465330 (98i:20039).

[12] R. Bieri, *Homological dimension of discrete groups*, Queen Mary College Mathematics Notes (1976). MR0466344 (57:6224).

[13] R. Bieri, *Normal subgroups in duality groups and in groups of cohomological dimension 2*, J. Pure Appl. Algebra 7 (1976) 35-51. MR0390078 (52:10904).

[14] J. S. Birman, A. Lubotzky and J. McCarthy, *Abelian and solvable subgroups of the mapping class groups*, Duke Math. J. **50** (1983) 1107-1120. MR726319 (85k:20126).

[15] N. Brady, M. Clay and P. Dani, *Morse theory and conjugacy classes of finite subgroups*, Geom. Dedicata **135** (2008) 15-22. MR2413325 (2009d:20103).

[16] M. R. Bridson, *On the subgroups of right angled Artin groups and mapping class groups*, Math. Res. Lett. **20** (2013) 1-10. MR3151642.

[17] M. R. Bridson, *Non-positive curvature and complexity for finitely presented groups*, in *Proceedings of the International Congress of Mathematicians, Madrid, 2006*, Vol. II (European Mathematical Society, 2006), pp.961-987. MR2275631 (2008a:20071).

[18] M.R. Bridson, *The rhombic dodecahedron and semisimple actions of $Aut(F_n)$ on $CAT(0)$ spaces*, Fund. Math. **214** (2011) 13-25. MR2845631 (2012h:20092).

[19] M. R. Bridson, *On the subgroups of semihyperbolic groups*, in E. Ghys et al., *Essays on Geometry and Related Topics: Mémoires Dédiés à André Haefliger* (Monographie no. 38 de l'Enseignement mathématique, 2001), pp. 85-111. MR1929323 (2003g:20068).

[20] M. R. Bridson, *Finiteness properties for subgroups of $GL(n, \mathbb{Z})$*, Math. Ann. **317** (2000) 629-633. MR1777113 (2001f:20112).

[21] M. R. Bridson, *Controlled embeddings into groups that have no non-trivial finite quotients*, in I. Rivin, C. Rourke and C. Series, eds., *The Epstein Birthday Schrift* (Geometry & Topology Monographs, Vol. 1, 1998), pp. 99-116. MR1668335 (2000g:20074).

[22] M. R. Bridson, *Decision problems and profinite completions of groups*, J. Algebra **326** (2011) 59-73. MR2746052 (2012f:20086).

[23] M. R. Bridson and F.J. Grunewald *Grothendieck's problems concerning profinite completions and representations of groups*, Ann. Math. **160** (2004) 359-373. MR2119723 (2005k:20069).

[24] M. R. Bridson and A. Haefliger, *Metric Spaces of Non-Positive Curvature* (Springer-Verlag, 1999). MR1744486 (2000k:53038).

[25] M. R. Bridson, J. Howie, C. F. Miller III and H. Short, *Subgroups of direct products of limit groups*. Ann. Math. **170** (2009) 1447-1467. MR2600879 (2011h:20082).

[26] M. R. Bridson, J. Howie, C. F. Miller III and H. Short, *On the finite presentation of subdirect products and the nature of residually free groups*. Am. J. Math. **135** (2013) 891-933. MR3086064.

[27] M. R. Bridson and C. F. Miller III, *Recognition of subgroups of direct products of hyperbolic groups*, Proc. Am. Math. Soc. **132** (2003) 59-65. MR2021248 (2004j:20068).

[28] M. R. Bridson and C. F. Miller III, *Structure and finiteness properties of subdirect products of groups*. Proc. Lond. Math. Soc. (3) **98** (2009) 631–651. MR2500867 (2010g:20049).

[29] M. R. Bridson and A. W. Reid, *Nilpotent completions of groups, Grothendieck pairs, and four problems of Baumslag*, Int. Math. Res. Not. **2015** (2015) 2111–2140. MR3344664.

[30] M. R. Bridson and R. D. Wade, Actions of higher rank lattices on free groups, Compositio Math. **147** (2011) 1573–1580. MR2834733 (2012j:20092).

[31] K. S. Brown, *Cohomology of Groups* (Springer-Verlag, 1982). MR672956 (83k:20002).

[32] P.-E. Caprace, *Lectures on proper CAT(0) spaces and their isometry groups*, in M. Bestvina, M. Sageev and K. Vogtman, eds., *Geometric Group Theory* (American Mathematical Society, 2014), pp. 91–126. MR3329726.

[33] M. Clay, C. Leininger and J. Mangahas, *The geometry of right angled Artin subgroups of mapping class groups*, Groups Geom. Dyn. **6** (2012) 249–278. MR2914860.

[34] D.J. Collins and C.F. Miller III, *The word problem in groups of cohomological dimension 2*, in C. M. Campbell, E. F. Robertson, N. Ruskuc and G. C. Smith, eds., *Groups St Andrews 1997 in Bath, I* (Cambridge University Press, 1999), pp. 211–218. MR1676618 (2000h:20061).

[35] J. Crisp and B. Wiest. *Quasi-isometrically embedded subgroups of braid and diffeomorphism groups*. Trans. Am. Math. Soc. **359** (2007) 5485–5503. MR2327038 (2008i:20048).

[36] F. Dahmani, V. Guirardel and D. Osin, *Hyperbolically embedded subgroups and rotating families in groups acting on hyperbolic spaces*, arXiv:1111.7048 [math.GT].

[37] M. W. Davis and T. Januszkiewicz, *Right-angled Artin groups are commensurable with right-angled Coxeter groups*, J. Pure Appl. Algebra **153** (2000) 229–235. MR1783167 (2001m:20056).

[38] C. Droms, *Graph Groups (Algebra, Kim, Roush, Magnus)* (PhD Thesis, Syracuse University, 1983). MR2633165.

[39] G. Duchamp and D. Krob, *The lower central series of the free partially commutative group*, Semigroup Forum **45** (1992) 385–394. MR1179860 (93e:20047).

[40] B. Farb, *Some problems on mapping class groups and moduli space*, Proc. Symp. Pure Math. **74** (2006) 11–56. MR2264130 (2007h:57018).

[41] B. Farb, A. Lubotzky and Y. Minsky, *Rank-1 phenomena for mapping class groups*, Duke Math. J. **106** (2001) 581–597. MR1813237 (2001k:20076).

[42] B. Farb and D. Margalit, *A Primer on Mapping Class Groups* (Princeton University Press, 2012). MR2850125 (2012h:57032).

[43] B. Farb and H. Masur, *Superrigidity and mapping class groups*, Topology **37** (1998) 1169–1176. MR1632912 (99f:57017).

[44] E.K. Grossman, *On the residual finiteness of certain mapping class groups*, J. Lond. Math. Soc. **9** (1974) 160–164. MR0405423 (53:9216).

[45] F. J. Grunewald, *On some groups which cannot be finitely presented*, J. Lond. Math Soc. **17** (1978) 427–436. MR500627 (80d:20033).

[46] F. Haglund, *Finite index subgroups of graph products*, Geom. Dedicata **135** (2008) 167–209. MR2413337 (2009d:20098).

[47] F. Haglund and D.T. Wise, *Special cube complexes*, Geom. Funct. Anal. **17** (2008) 1551–1620. MR2377497 (2009a:20061).

[48] G. Higman, *A finitely generated infinite simple group*, J. Lond. Math Soc. **26** (1951) 61–64. MR0038348 (12:390c).

[49] S. P. Humphries, *On representations of Artin groups and the Tits conjecture*, J. Algebra **169** (1994) 847–862. MR1302120 (95k:20057).

[50] N. Ivanov, *Subgroups of Teichmüller Modular Groups* (Translations of Mathematical Monographs, Vol. 115, American Mathematical Society). MR1195787 (93k:57031).

[51] S.P. Kerckhoff, *The Nielsen realization problem*, Ann. Math. **117** (1983) 235–265. MR690845 (85e:32029).

[52] T. Koberda. *Right-angled Artin groups and a generalized isomorphism problem for finitely generated subgroups of mapping class groups*, Geom. Funct. Anal. **22** (2012) 1541–1590. MR3000498.

[53] R. Kulkarni, *Symmetries of surfaces*, Topology **26** (1987) 195–203. MR895571 (88m:57051).

[54] J. McCarthy, *A 'Tits-alternative' for subgroups of surface mapping class groups*, Trans. Am. Math. Soc. **291** (1985) 583–612. MR800253 (87f:57011).

[55] C. F. Miller III, *On Group-Theoretic Decision Problems and their Classification* (Princeton University Press, 1971). MR0310044 (46:9147).

[56] C. F. Miller III, *Decision problems for groups: survey and reflections*, in C. Baumslag and C. F. Miller III, eds., *Algorithms and Classification in Combinatorial Group Theory* (MSRI/Springer-Verlag, 1992), pp. 1–59. MR1230627 (94i:20057).

[57] K. A. Mihailova, *The occurrence problem for direct products of groups* [in Russian], Dokl. Akad. Nauk SSSR **119** (1958) 1103–1105. MR0100018 (20:6454).

[58] A. Minasyan, *Hereditary conjugacy separability of right angled Artin groups and its applications*, Groups Geom. Dyn. **6** (2012) 335–388. MR2914863.

[59] G. Mislin, *Classifying spaces for proper actions of mapping class groups*, Münster J. Math. **3** (2010) 263–272. MR2775365 (2012c:55015).

[60] L. Mosher, *Mapping class groups are automatic*, Ann. Math. **142** (1995) 303–384. MR1343324 (96e:57002).

[61] V. P. Platonov and O. I. Tavgen, *Grothendieck's problem on profinite completions and representations of groups*, K-Theory **4** (1990) 89–101. MR1076526 (93a:20040).

[62] E. Rips, *Subgroups of small cancellation groups*, Bull. Lond. Math. Soc. **14** (1982) 45–47. MR642423 (83c:20049).

[63] M. Sageev, *CAT(0) cube complexes and groups*, in M. Bestvina, M. Sageev and K. Vogtman, eds., *Geometric Group Theory* (American Mathematical Society, 2014), pp. 7–54. MR3329724.

[64] J. R. Stallings, *Homology and central series of groups*, J. Algebra **2** (1965) 170–181. MR0175956 (31:232).

[65] J. R. Stallings, *A finitely presented group whose 3-dimensional integral homology is not finitely generated*, Am. J. Math. **85** (1963) 541–543. MR0158917 (28:2139).

[66] R. D. Wade, *Johnson homomorphisms and actions of higher-rank lattices on right-angled Artin groups*, J. Lond. Math. Soc. **88** (2013) 860–882. MR3145135.

[67] R. D. Wade, *The lower central series of a right-angled Artin group*, Enseign. Math. **61** (2015) 343–371. MR3539842.

[68] D. T. Wise, *Cubulating small cancellation groups*, Geom. Funct. Anal. **14** (2004) 150–214. MR2053602 (2005c:20069).

[69] D. T. Wise, *The structure of groups with a quasiconvex hierarchy*, Preprint, McGill University (2011).

4 Polyfolds and Fredholm Theory

HELMUT H. W. HOFER

4.1 Introduction

In this chapter, we discuss the generalized Fredholm theory in polyfolds. The initial version of the chapter, [23], was written in 2008 on the occasion of a lecture given at the Clay Mathematical Institute (CMI) and described the theory as developed in [25–27]. Six years is a long time, and since then the theory has advanced considerably—a comprehensive discussion of its more abstract parts can be found in [31, 32]. Currently non-trivial applications are being developed, most notably those to symplectic field theory (SFT). In order to account for these developments, the chapter has been to a large extent rewritten. Rather than dealing with the general theory, which also allows for a boundary with corners, we restrict ourselves to a special case and illustrate it with a discussion of stable maps, a topic closely related to Gromov–Witten theory. We also would like to mention the paper [11], where the ideas of polyfold theory are explained as well. The abstract theory has been applied in [30] as part of the general construction of SFT. In [57], it was used to address the Weinstein conjecture in higher dimensions. An extensive study of the case with boundary and corners is contained in [31, 32]. These extensions are crucial for applications to SFT. A basic paper in this direction is [12], in which the polyfolds relevant for SFT are constructed. This is the place where the full power of the theory becomes apparent.

We shall start by introducing the category of stable maps. The construction of Gromov–Witten invariants can be understood as the geometric study of perturbations of the full subcategory of J-holomorphic stable maps. We take a more general viewpoint and study the whole category. Our initial discussion is entirely topological, and surprisingly the objects that arise are so natural that one is forced to raise the question of whether these structures go beyond topology. Indeed they do, and they give rise to a generalized differential geometry as well as a generalized nonlinear Fredholm theory accompanied by its own blend of nonlinear analysis. The whole package is referred to as polyfold

Lectures on Geometry. Edward Witten, Marc Lackenby, Martin R. Bridson, Helmut Hofer and Rahul Pandharipande.
© Oxford University Press 2017. Published 2017 by Oxford University Press.

theory. This polyfold theory provides a language and a large body of results to address questions arising in the study of moduli problems in symplectic geometry. It is clear that its applicability goes far beyond the latter field. From a nonlinear analysis perspective, the symplectic applications are concerned with the study of isomorphism classes of families of nonlinear first-order elliptic equations with varying domains and varying targets. The domains are even allowed to change the topology, and bubbling-off phenomena will in general occur. This is the source for compactness and transversality issues, which make an algebraic counting of solutions very difficult. In the case of the problems arising in symplectic geometry, the polyfold theory overcomes these difficulties. There is no doubt that the theory should have applications in other parts of nonlinear analysis as well.

It is a basic observation in symplectic geometry/topology that geometric questions can be rephrased as questions about solution spaces of nonlinear first-order elliptic partial differential equations (the solution spaces are by definition the moduli spaces). Very often, just counting solutions suffices to answer a seemingly difficult geometric question. The construction of the moduli problems are themselves difficult analytical problems, involving analytical limiting behaviors, like bubbling-off, breaking of trajectories and stretching of the neck. The naive solution sets are very often not compact and are not cut out by the differential equation in a generic way. In most applications, there exist intriguing compactifications of the solution spaces, which, however, are usually not compatible with the standard versions of smooth nonlinear analysis based on the notion of Fréchet differentiability. Moreover, as a consequence of local symmetries, very often even a generic choice of geometric auxiliary data used to construct the partial differential equation will never result in a generic solution set. The polyfold theory allows one to view the partial differential equation within an abstract framework, which provides a Sard–Smale perturbation theory and deals with the transversality issues. The framework is so general that it also encompasses the geometric perturbation theory. Hence, one can proceed geometrically as long as this is possible, and use the abstract perturbations only if the problems cannot be dealt with geometrically. In particular, whatever has been established to work classically will also work in this extended framework. The papers [34, 50, 60] give an idea of what is classically possibly. If one wants to go further, then one needs to incorporate abstract perturbations.

The analytical limiting phenomena that we mentioned above, even assuming a sufficient amount of genericity, do not look like smooth phenomena if smoothness refers to the usual concept. Indeed, the coordinate changes (for the ambient spaces containing the moduli spaces) are from a classical perspective usually nowhere-differentiable. However, it turns out that the notion of smoothness can be relaxed, and a generalization of differential geometry and nonlinear functional analysis can be developed, for which the limiting phenomena can be viewed as smooth phenomena, even if they are quite often obscured by transversality issues. In this generalized context, the classical nonlinear Fredholm theory can be extended to a much larger class of spaces and operators, which can deal with the aforementioned problems.

A great example to explain the theory comes from a study of the category of stable maps associated to a symplectic manifold. The study of these objects goes back to the seminal work by Gromov on pseudoholomorphic curve theory—see [18]. Gromov showed that the study of pseudoholomorphic maps is a powerful tool in symplectic geometry, demonstrating its uses by many examples. Kontsevich [36] later pointed out the importance of the notion of stable pseudoholomorphic curves. Our stable maps need not to be pseudoholomorphic. We find several natural constructions that at first glance are just of topological nature without exhibiting more regularity. However, as it turns out, these are the shadows of smooth constructions, once the notion of differentiability in finite dimensions, which is usually generalized as Fréchet differentiability to infinite-dimensional Banach spaces, is generalized in a quite different way. Such a generalization requires an additional piece of structure, called an sc-structure, which occurs in inter-polation theory [59], albeit under a different name. In fact, we give a quite different interpretation of such a structure and make clear that it can be viewed as a generalization of a smooth structure on a Banach space. We call this generalization sc-smoothness, and the generalization of differentiability of a map we refer to as sc-differentiability. The use of scales in a geometric nonlinear analysis setting is not new—see the work by D. Ebin [7] and H. Omori, [40] (these references were pointed out by T. Mrowka). The ideas in the latter paper are somewhat closer to our viewpoint. However, Omori does not use com-pact scales, and this turns out to be a crucial condition for our applications. For example, without the compactness assumption, one does not obtain the new local models for a generalized differential geometry. This we discuss next.

The very interesting aspect is then the following fact. There are many sc-differentiable maps $r : U \to U$ satisfying $r \circ r = r$, i.e., sc-smooth retractions. For Fréchet differentiab-ility, the image of such a retraction can be shown to be a submanifold of U. Incidentally, this is H. Cartan's last mathematical theorem—see [4] (this reference was pointed out by E. Ghys). However, the images of sc-smooth retractions can be much more general. Most strikingly, they can have locally varying dimensions. Of course, a good notion of differ-entiability comes with the chain rule, so that from $r \circ r = r$ we deduce $Tr = (Tr) \circ (Tr)$. In other words, the tangent map of an sc-smooth retraction is again an sc-smooth retrac-tion. If a subset O of an sc-Banach space is the image of an sc-smooth retraction r, then $TO = Tr(TU)$ defines the tangent space, and it turns out that the definition does not depend on the choice of r as long as r is sc-smooth and has O as its image. So we obtain quite general subsets of Banach spaces that have tangent spaces. An sc-smooth map $f : O \to O'$, where $O \subset E$ and $O' \subset F$ are sc-smooth retracts, is a map such that $f \circ r : U \to F$ is sc-smooth, where r is an sc-smooth retraction onto O. As it turns out, the defin-ition does not depend on the choice of r. Further, one verifies that $Tf := T(f \circ r)|TO$ defines a map $TO \to TO'$ between tangent spaces and that the definition also does not depend on the choice of r.

In summary, once we have a good notion of differentiability for maps between open sets, we also obtain a notion of differentiability for maps between sc-smooth retracts. However, for the usual notion of differentiability, smooth retracts are manifolds and one does not obtain anything beyond the usual differential geometry and its standard gen-eralization to Banach manifolds. On the other hand, sc-differentiability opens up new possibilities with serious applications. We generalize differential geometry by general-izing the notion of a manifold to that of an M-polyfold. These are metrizable spaces that are locally homeomorphic to retracts with sc-smooth transition maps. The theory

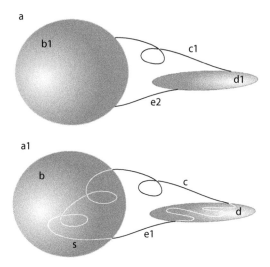

Figure 4.1 (a) A finite-dimensional M-polyfold X that is homeomorphic to the space consisting of the disjoint union of an open three-ball B^3 and an open two-ball B^2 connected by two curves c_1 and c_2. (b) The same M-polyfold containing a one-dimensional S^1-like submanifold S. This submanifold could arise as the zero-set of a transversal section of a strong M-polyfold bundle Y over X, which has varying dimensions. Namely, over the three-ball it is two-dimensional, over the two-disk it is one-dimensional and otherwise it is trivial. The polyfold theory would then guarantee a natural smooth structure on the solution set S.

as described in this chapter even gives new objects in finite dimensions (Fig. 4.1). Most important for us is the fact that the new local models for a vastly generalized differential geometry allow us to build spaces, which can be used to bring nonlinear partial differential equations that might show bubbling-off phenomena into an abstract geometric framework that allows for a very general nonlinear Fredholm theory with the usual expected properties.

4.2 A motivating example and natural structures

The following discussion starts with a closer look at the category of stable Riemann surfaces and describes some of the interesting structures appearing as a result of the classical Deligne–Mumford theory [6]. These structures and the associated viewpoint can be related to a topic arising in Gromov–Witten theory, namely, the study of the category of stable maps that are not necessarily pseudoholomorphic. This category of stable maps has natural structures. Natural structures in mathematics are usually there for a good reason. In our case, a deeper analysis reveals the existence of more general smooth models for an extended differential geometry as well as a vastly generalized Fredholm theory. The latter can be used to define the Gromov–Witten invariants. The same scheme also works suitably extended in the much more complicated framework of stable maps in SFT.

4.2.1 The category of stable noded Riemann surfaces

We follow the presentation in [24], which is somewhat different from that given in [53]. A good starting point for explaining some of the later features in the polyfold theory is a study of stable noded Riemann surfaces with marked points.

Definition 4.1. *A stable, possibly noded Riemann surface with marked points is a tuple* (S, j, M, D), *where* (S, j) *is a closed smooth Riemann surface,* $M \subset S$ *is a finite collection of points, called marked points, and* D *is a finite collection of unordered pairs* $\{x, y\}$ *of points* $x, y \in S \setminus M$ *having the following properties:*

(i) *For* $\{x, y\} \in D$, *it holds that* $x \neq y$. *If* $\{x, y\} \cap \{x', y'\} \neq \varnothing$, *then* $\{x, y\} = \{x', y'\}$.

(ii) *The topological space* $\bar{S}$ *obtained by identifying* $x \equiv y$ *for* $\{x, y\} \in D$ *is connected.*

(iii) *Define the subset* $|D|$ *by* $|D| = \bigcup_{\{x,y\} \in D} \{x, y\}$. *For every connected component* C *of* S *having genus* $g(C)$, *with* n_C *being the number of points in* $C \cap (|D| \cup M)$, *it holds that* $2g(C) + n_C \geq 3$.

Condition (iii) is called the stability condition. In our case, the set M is not ordered, but the following discussion applies to the ordered case as well. We can view $\alpha = (S, j, M, D)$ as an object in a category $\mathcal{R}$, where the morphisms $\Phi : \alpha \to \alpha'$ are given by

$$\Phi = (\alpha, \phi, \alpha') : \alpha \to \alpha'$$

and $\phi : (S, j) \to (S', j')$ is a biholomorphic map satisfying $\phi(M) = M'$ and $\phi(D) = D'$, where $\phi(D) := \{\{\phi(x), \phi(y)\} \mid \{x, y\} \in D\}$.

Definition 4.2. *We call* $\mathcal{R}$ *the category of stable, (possibly) noded Riemann surfaces with unordered marked points.*

The following result is well known, and is part of the Deligne–Mumford theory. The theory has been described from a more differential geometric perspective in [53], and is described in [24] from a polyfold perspective.

Theorem 4.3. *The category* $\mathcal{R}$ *has the following properties:*

(i) *Every morphism in the category* $\mathcal{R}$ *is an isomorphism.*

(ii) *The stability condition implies that between any two objects there are at most finitely many morphisms.*

(iii) *The orbit space* $|\mathcal{R}|$ *of the category* $\mathcal{R}$, *which is the set of isomorphism classes of objects in* $\mathcal{R}$, *carries a natural metrizable topology.*

(iv) *The topological space* $|\mathcal{R}|$ *carries in a natural way the structure of a holomorphic orbifold for which each connected component is compact.*

A basis for the topology is given in Proposition 4.11. We shall need a variation of this result, and shall describe a particular approach to $\mathcal{R}$ that one might view as a toy case for the polyfold approach to stable maps in symplectic manifolds. We describe some of the results we shall need later on, and refer the reader for more details to $[24, 30]$ or the references mentioned therein.

Given an object $\alpha = (S, j, M, D)$ in $\mathcal{R}$, it is not just a point in the category, but has geometry as well. In fact, the additional structures on the category come from the fact that every object has its own geometry, which is being exploited by constructing associated objects by a plumbing procedure. Recall that the automorphism group G of α is the finite collection of all morphisms (which are all isomorphisms) $\Phi : \alpha \to \alpha$.

Definition 4.4. *Let $\alpha = (S, j, M, D)$ be an object in $\mathcal{R}$ with automorphism group G. A small disk structure associated to α and denoted by $\mathbf{D}$ assigns to every $x \in |D|$ a compact disk-like neighborhood $D_x \subset S$ around the point x, so that the following hold:*

   (i) D_x has a smooth boundary and the disks are mutually disjoint.

   (ii) $\bigcup_{x \in |D|} D_x$ is invariant under G.

   (iii) $D_x \cap M = \varnothing$ for all $x \in |D|$.

Given a small disk structure $\mathbf{D}$ we have for every $\{x, y\} \in D$ an associated noded disk pair $(D_x \cup D_y, \{x, y\})$. From this data, we can construct for every $\{x, y\}$ a natural gluing parameter. This is done as follows. Denote for $x \in |D|$ by $\widehat{x}$ an oriented real line in $(T_x S, j)$. If $\mathbb{S}^1 \subset \mathbb{C}$ denotes the unit circle, then we observe that given $\theta \in \mathbb{S}^1$, we can define a new oriented real line $\theta \widehat{x}$, making use of the complex structure j. We introduce an equivalence class of unordered pairs of real oriented lines $\{\widehat{x}, \widehat{y}\}$ associated to $\{x, y\} \in D$ as follows. We say $\{\widehat{x}, \widehat{y}\}$ is equivalent to $\{\widehat{x}', \widehat{y}'\}$ (here the lines lie over x and y) provided there exists $\theta \in \mathbb{S}^1$ satisfying

$$\widehat{x} = \theta \widehat{x}' \quad \text{and} \quad \widehat{y} = \theta^{-1} \widehat{y}'.$$

Equivalence classes will be written as $[\widehat{x}, \widehat{y}]$. Clearly, the collection of all equivalence classes associated to the pair $\{x, y\}$ can be parametrized by the unit circle via

$$\mathbb{S}^1 \ni \theta \to [\widehat{x}, \theta \widehat{y}].$$

Definition 4.5. *An equivalence class $[\widehat{x}, \widehat{y}]$ is called a decorated nodal pair with underlying nodal pair $\{x, y\}$.*

Hence, above every $\{x, y\} \in D$ there lies an $\mathbb{S}^1$-worth of decorated nodal pairs $[\widehat{x}, \widehat{y}]$. We shall refer to $[\widehat{x}, \widehat{y}]$ as a decoration for $\{x, y\}$.

Consider a formal expression $r[\widehat{x}, \widehat{y}]$ with $r \in [0, \frac{1}{2})$. Two such formal expressions are considered to be the same, i.e., $r[\widehat{x}, \widehat{y}] = r'[\widehat{x}', \widehat{y}']$, provided one of the following holds:

   (i) $r = r' = 0$, or

   (ii) $r = r' \neq 0$ and $[\widehat{x}, \widehat{y}] = [\widehat{x}', \widehat{y}']$.

Definition 4.6. *A natural gluing parameter associated to the nodal pair $\{x, y\}$ is a formal expression $\mathfrak{a}^{\{x,y\}} = r[\widehat{x}, \widehat{y}]$ where $r \in [0, \frac{1}{2})$, with the notion of equality as defined above. In the case $r = 0$, we shall write $\mathfrak{a}^{\{x,y\}} = 0$.*

We define the set $\mathbb{B}^{\{x,y\}}(\frac{1}{2})$ by

$$\mathbb{B}^{\{x,y\}}(\tfrac{1}{2}) = \{r[\widehat{x},\widehat{y}] \mid r \in [0,\tfrac{1}{2}), \ [\widehat{x},\widehat{y}] \text{ a decoration of } \{x,y\}\}$$

and call it the set of natural gluing parameters associated to the nodal pair $\{x,y\}$. The modulus of $\mathfrak{a}^{\{x,y\}} = r[\widehat{x},\widehat{y}]$ is the number r and we shall write $|\mathfrak{a}^{\{x,y\}}| = r$.

Definition 4.7. *Given α, a natural (total) gluing parameter is a map*

$$\mathfrak{a} : D \ni \{x,y\} \to \mathfrak{a}^{\{x,y\}} \in \mathbb{B}^{\{x,y\}}(\tfrac{1}{2}).$$

The collection of all gluing parameters associated to α will be denoted by $\mathbb{B}^\alpha$.

The dependence of $\mathbb{B}^\alpha$ on α is as follows. It depends only on the complex multiplication on the tangent spaces $T_x S$ for $x \in |D|$. We note the following result, which has an easy proof that is left to the reader.

Proposition 4.8. *Assume that α is given. For every $\{x,y\} \in D$, the set $\mathbb{B}^{\{x,y\}}(\frac{1}{2})$ has in a natural way the structure of a holomorphic manifold. It is characterized by the following property. For every choice of oriented real lines $\widehat{x}$ and $\widehat{y}$, the map*

$$\mathbb{B}^{\{x,y\}}(\tfrac{1}{2}) \to \{z \in \mathbb{C} \mid |z| < \tfrac{1}{2}\} : r[\widehat{x},\theta\widehat{y}] \to r\theta$$

is biholomorphic. In particular, the set $\mathbb{B}^\alpha$ of all natural gluing parameters $\mathfrak{a} : \{x,y\} \to \mathfrak{a}^{\{x,y\}}$ has a natural holomorphic structure and is biholomorphic to a product of open disks.

We shall use the natural gluing parameters to modify α and produce new stable noded Riemann surfaces. In this context, an important concept is that of a gluing profile.

Definition 4.9. *A gluing profile is a smooth diffeomorphism $\varphi : (0,1] \to [0,\infty)$.*

We shall refer to

$$\varphi(r) = -\frac{1}{2\pi} \ln r$$

as the logarithmic gluing profile and to

$$\varphi(r) = e^{1/r} - e$$

as the exponential gluing profile. Restricting $r[\widehat{x},\widehat{y}]$ to the case with $r \in [0,\frac{1}{2})$ has to do with the choice of these two gluing profiles. For an arbitrary gluing profile φ, we only have to require that $r \in [0,\varepsilon_\varphi)$, where $\varepsilon_\varphi > 0$, depending on the profile, is small enough.

Assume we are given a small disk structure $\mathbf{D}$ and a gluing profile φ. We can define a gluing or plumbing construction as follows. For a gluing parameter $\mathfrak{a}^{\{x,y\}} = r[\widehat{x},\widehat{y}]$ associated to $\{x,y\} \in D$, we consider the disks D_x and D_y and the nodal pair $\{x,y\}$. If $\mathfrak{a}^{\{x,y\}} = 0$, we keep this data, i.e., in other words, the gluing of $(D_x \cup D_y, \{x,y\})$ with $\mathfrak{a}^{\{x,y\}} = 0$ produces the noded disk $(D_x \cup D_y, \{x,y\})$. If $\mathfrak{a}^{\{x,y\}} \neq 0$, we compute the value $R = \varphi(r)$

and take a representative $\{\widehat{x}, \widehat{y}\}$ of the class $[\widehat{x}, \widehat{y}]$. After this choice, there exist uniquely determined biholomorphic maps

$$h_x : (D_x, x) \to (\mathbb{D}, 0), \quad Th_x(x)\widehat{x} = \mathbb{R},$$

and

$$h_y : (D_y, y) \to (\mathbb{D}, 0), \quad Th_y(x)\widehat{y} = \mathbb{R}.$$

Here $\mathbb{D}$ is the closed unit disk in $\mathbb{C}$, and $\mathbb{R} \subset T_0\mathbb{D}$ is oriented by 1. Recall that the modulus of the annulus $A_{r_1, r_2} := \{z \in \mathbb{C} \mid r_1 \leq |z| \leq r_2\}$ is defined by

$$\text{modulus}(A_{r_1, r_2}) = \frac{1}{2\pi} \ln\left(\frac{r_2}{r_1}\right).$$

If we are given a Riemann surface (A, j), where A is compact, has smooth boundary components, and is diffeomorphic to an annulus, then we know from the uniformization theorem that (A, j) is biholomorphic to $A_{r,1}$ for a uniquely determined $0 < r < 1$. The modulus of (A, j) is, by definition, given by

$$\text{modulus}(A, j) = -\frac{1}{2\pi} \ln r.$$

We note that the right-hand side is precisely $R = \varphi(r)$ for the logarithmic gluing profile. In the next step, the choice of the gluing profile matters, and a remark, why we consider other gluing profiles than the logarithmic one, is in order.

Remark 4.10. In later constructions, we shall see that the sc-smooth structure we are going to construct will depend on the choice of φ. The sc-smooth structure also influences if certain operators are sc-smooth Fredholm operators. The operators we are interested in will be sc-smooth for the exponential gluing profile, but not for the logarithmic one.

With some gluing profile φ fixed, we pick concentric compact annuli $A_x \subset D_x$ and $A_y \subset D_y$ with one boundary component being the boundary of D_x and D_y, respectively, so that A_x and A_y have modulus $R = \varphi(r)$. We discard the points $D_x \setminus A_x$ and $D_y \setminus A_y$ from S and the nodal pair $\{x, y\}$. With $A_x \cup A_y \subset S \setminus ((D_x \setminus A_x) \cup (D_y \setminus A_y))$, we identify A_x with A_y by identifying $z \equiv z'$, where $(z, z') \in A_x \times A_y$ satisfying

$$h_x(z)h_y(z') = e^{-2\pi R}.$$

The definition of the identification $z \equiv z'$ is independent of the choice of the representatives $\{\widehat{x}, \widehat{y}\}$ in $[\widehat{x}, \widehat{y}]$. Let us denote by $\mathcal{Z}_{a^{\{x,y\}}}$ the glued space obtained by identifying the two annuli. If $a^{\{x,y\}} = 0$, this is the noded $(D_x \cup D_y, \{x, y\})$ and for $a^{\{x,y\}} \neq 0$ it is the space obtained by identifications just described. Clearly, in the latter case, we have natural biholomorphic maps

$$A_x \to \mathcal{Z}_{\mathfrak{a}^{\{x,y\}}} \quad \text{and} \quad A_y \to \mathcal{Z}_{\mathfrak{a}^{\{x,y\}}},$$

by associating to a point its equivalence class. Carrying out the above procedure for every noded disk pair $(D_x \cup D_y, \{x,y\})$ defines for a gluing parameter $\mathfrak{a} \in \mathbb{B}^\alpha$ a new noded stable Riemann surface $(S_\mathfrak{a}, j_\mathfrak{a}, M_\mathfrak{a}, D_\mathfrak{a})$ with marked points. We have just described how to obtain $S_\mathfrak{a}$. We denote by $j_\mathfrak{a}$ the natural smooth almost complex structure on the latter. Observe that $z \equiv z'$ is a holomorphic identification. By $M_\mathfrak{a}$ we just mean M naturally identified with a subset of $S_\mathfrak{a}$. The collection of nodal points $D_\mathfrak{a}$ is obtained by first removing from D all $\{x,y\}$ with $\mathfrak{a}^{\{x,y\}} \neq 0$ and then identifying this set with the obvious set of nodal pairs for $S_\mathfrak{a}$. For every $\{x,y\} \in D_\mathfrak{a}$, we still have $D_x \cup D_y \subset S_\mathfrak{a}$.

At this point, we can exhibit a basis of open neighborhoods for a given point $|\alpha| \in |\mathcal{R}|$ for the natural metrizable topology. For a representative α and an associated gluing parameter $\mathfrak{a}$, we define $|\mathfrak{a}|$ by

$$|\mathfrak{a}| = \max_{\{x,y\} \in D} \left| \mathfrak{a}^{\{x,y\}} \right|.$$

Write $\alpha = (S, j, M, D)$ and fix a small disk structure $\mathbf{D}$. The set of all smooth almost complex structures k on S, which define the same orientation as j, has a natural metrizable topology measuring the C^∞-distance. Denote a choice of metric by ρ. For a given gluing profile φ and $\varepsilon > 0$ small enough, define

$$\mathcal{U}(\alpha, \mathbf{D}, \varphi, \varepsilon) = \{|(S_\mathfrak{a}, k_\mathfrak{a}, M_\mathfrak{a}, D_\mathfrak{a})| \mid \rho(j,k) < \varepsilon, \, j = k \text{ on } \mathbf{D}, \, |\mathfrak{a}| < \varepsilon\}.$$

Theorem 4.11. *Fix a gluing profile φ, pick for every class $c \in |\mathcal{R}|$ an object α_c and choose an associated small disk structure $\mathbf{D}_c$ for α_c. Consider $\mathcal{U}(\alpha_c, \mathbf{D}_c, \varphi, \varepsilon)$ for $0 < \varepsilon < \varepsilon_c$ and denote this collection by $\mathfrak{U}_c$. Then the union $\mathfrak{U} = \bigcup_{c \in |\mathcal{R}|} \mathfrak{U}_c$ is the basis for a metrizable topology $\mathcal{T}$ on $|\mathcal{R}|$. The topology $\mathcal{T}$ does not depend on the choices involved in constructing $\mathfrak{U}$.*

The topology $\mathcal{T}$ is the one referred to in Theorem 4.3. The automorphism group G acts on the set of gluing parameters $\mathbb{B}^\alpha$ via biholomorphic maps

$$G \times \mathbb{B}^\alpha \to \mathbb{B}^\alpha : (g, \mathfrak{a}) \to g * \mathfrak{a}$$

in a natural way as follows. Namely, $\mathfrak{b} = g * \mathfrak{a}$ is defined by

$$\mathfrak{b}^{\{g(x), g(y)\}} = r[Tg(x)\hat{x}, Tg(y)\hat{y}],$$

where $r[\hat{x}, \hat{y}] = \mathfrak{a}^{\{x,y\}}$. With the above construction, an element $g \in G$ induces a biholomorphic map

$$g_\mathfrak{a} : (S_\mathfrak{a}, j_\mathfrak{a}, M_\mathfrak{a}, D_\mathfrak{a}) \to (S_\mathfrak{b}, j_\mathfrak{b}, M_\mathfrak{b}, D_\mathfrak{b}),$$

where $\mathfrak{b} = g * \mathfrak{a}$. It also holds that

$$h_{g*\mathfrak{a}} \circ g_\mathfrak{a} = (h \circ g)_\mathfrak{a}.$$

Let us abbreviate $(S_\mathfrak{a}, j_\mathfrak{a}, M_\mathfrak{a}, D_\mathfrak{a})$ by $\alpha_\mathfrak{a}$. Then $\alpha_\mathfrak{a} = \alpha_{\mathfrak{a}'}$ if and only if $\mathfrak{a} = \mathfrak{a}'$. Consider for these objects $\alpha_\mathfrak{a}$ with $|\mathfrak{a}| < \frac{1}{2}$ the associated full subcategory $\mathcal{A} = \mathcal{A}_{\alpha, D, \varphi}$ of $\mathcal{R}$. This subcategory is small. Given the object α with automorphism group G, every choice of a small disk structure (a set of possible choices) produces such a subcategory. We also consider the translation groupoid $G \ltimes \mathbb{B}^\alpha$. The latter is a small category with object set $\mathbb{B}^\alpha$ and morphism set $G \times \mathbb{B}^\alpha$, where $(g, \mathfrak{a})$ is seen as a morphism $\mathfrak{a} \to g * \mathfrak{a}$. Observe that object set and morphism set have in our case both smooth manifold structures. There exists the obvious functor

$$\gamma : G \ltimes \mathbb{B}^\alpha \to \mathcal{A},$$

which maps the object $\mathfrak{a}$ to $\alpha_\mathfrak{a}$ and the morphism $(g, \mathfrak{a}) : \mathfrak{a} \to g * \mathfrak{a}$ to

$$(\alpha_\mathfrak{a}, g_\mathfrak{a}, \alpha_{g * \mathfrak{a}}) : \alpha_\mathfrak{a} \to \alpha_{g * \mathfrak{a}}.$$

We note that γ is injective on objects. More is true, namely, from the results of the Deligne–Mumford theory, we obtain the following proposition, which is true for any gluing profile. It has to be viewed as an intermediate result neglecting for the moment the possibility of deforming the (integrable) almost complex structure j associated to α.

Proposition 4.12. *For a suitable G-invariant open neighborhood U of $0 \in \mathbb{B}^\alpha$ depending on the choice of the gluing profile, the functor*

$$\gamma : G \ltimes U \to \mathcal{R}$$

is injective on objects and fully faithful.

The above proposition describes the family we can obtain just from gluing (or plumbing) without considering deformations of the (integrable) smooth almost complex structure. We can also obtain from α new objects by changing the complex structure. Of course, a priori, we only would like to consider those deformations that produce non-isomorphic objects. There is, however, an inherent difficulty coming from the automorphisms. Given the object α in the category $\mathcal{R}$, consider the complex vector space $\Gamma_0(\alpha)$ consisting of smooth sections of $TS \to S$ that vanish at the points in $M \cup |D|$. By $\Omega^{0,1}(\alpha) \to S$ we denote the complex vector space of smooth sections of $\text{Hom}_\mathbb{R}(TS, TS) \to S$ which are complex antilinear. The Cauchy–Riemann operator defines a linear operator

$$\bar{\partial} : \Gamma_0(\alpha) \to \Omega^{0,1}(\alpha).$$

Define the arithmetic genus via the formula

$$g_a := 1 + \sharp D + \sum_C [g(C) - 1],$$

where the sum is taken over all connected components C of S. The basic fact is given by the following proposition.

Proposition 4.13. *The Cauchy–Riemann operator is a complex linear differential operator. The stability property of α implies that $\bar{\partial}$ is injective and as a consequence of Riemann–Roch the cokernel has complex dimension $3g_a + \sharp M - \sharp D - 3$.*

In the next step, we consider the deformation space of α for fixed combinatorial type. This means we consider the deformation of the complex structure, the marked points and the nodal pairs, but we do not remove nodes. The natural deformation space introduced in the following is the infinitesimal version of this, if we divide out by isomorphisms but keep track of the automorphisms.

Definition 4.14. *The (infinitesimal) natural deformation space (of fixed combinatorial type) of $\alpha = (S, j, M, D)$ is by definition the complex vector space $H^1(\alpha)$ defined by*

$$H^1(\alpha) = \Omega^{0,1}/(\bar{\partial}\Gamma_0(\alpha)).$$

The construction $\alpha \to H^1(\alpha)$ is functorial. Namely, we associate to an object α a finite-dimensional complex vector space and to a morphism $\phi : \alpha \to \alpha'$ a complex vector space isomorphism

$$H^1(\phi) : H^1(\alpha) \to H^1(\alpha') : [\tau] \to [\phi \circ \tau \circ \phi^{-1}].$$

In particular, the automorphism group G of the object α has an action on $H^1(\alpha)$.

Assume that V is an open subset of some vector space E and $V \in v \to j(v)$ a smooth family of almost complex structures on S. Then it holds that $j(v)^2 = -Id$, and differentiating at v in the direction $\delta v \in E$ gives the identity

$$j(v)(Dj(v)\delta v) = -(Dj(v)\delta v)j(v).$$

This means that for every $z \in S$, the map

$$(Dj(v)\delta v)_z : (T_z S, j(v)) \to (T_z S, j(v))$$

is complex antilinear and induces an element $[Dj(v)\delta v] \in H^1(S, j(v), M, D)$. Hence we obtain for every $v \in V$ a linear map

$$[Dj(v)] : E \to H^1(S, j(v), M, D).$$

This map is called the Kodaira–Spencer differential associated to $v \to j(v)$. We define α_v by

$$\alpha_v = (S, j(v), M, D).$$

Definition 4.15. *Given a stable α and an associated small disk structure D, we shall call a smooth family $v \to j(v)$ of almost complex structures on S defined on a G-invariant open neighborhood V of 0 in $H^1(\alpha)$ a good deformation compatible with D provided it has the following properties:*

(i) $j(0) = j$.

(ii) $j(v) = j$ on all disks D_x associated to **D**.

(iii) For every $v \in V$, the Kodaira–Spencer differential

$$[Dj(v)] : H^1(\alpha) \to H^1(\alpha_v)$$

is a complex linear isomorphism.

(iv) For every $g \in G$ and $v \in V$, the map $g : \alpha_v \to \alpha_{g*v}$ is biholomorphic.

We note that $v \to j(v)$ for small v is necessarily an injective map into the space of smooth almost complex structures.

A well-known result with a proof given in [24] is the following.

Theorem 4.16. *Given a stable α in $\mathcal{R}$ with automorphism group G, and a small disk structure* **D***, there always exist good deformations compatible with* **D***.*

Suppose we start with a stable α with automorphism group G and have fixed a small disk structure **D**. Then we can, as we have seen, construct the functor

$$\gamma : G \ltimes \mathbb{B}^\alpha \to \mathcal{R}.$$

Now taking a good deformation as described in Definition 4.15, say $V \ni v \to j(v)$, we can construct the following functor, where we take either the logarithmic or exponential gluing profile:

$$\Psi : G \ltimes (V \times \mathbb{B}^\alpha) \to \mathcal{R}.$$

To objects $(v, \mathfrak{a})$ we assign

$$\alpha_{(v,\mathfrak{a})} = (S_\mathfrak{a}, j(v)_\mathfrak{a}, M_\mathfrak{a}, D_\mathfrak{a}).$$

The morphism $(g, (v, \mathfrak{a})) : (v, \mathfrak{a}) \to (g * v, g * \mathfrak{a})$ is mapped to

$$(\alpha_{(v,\mathfrak{a})}, g_\mathfrak{a}, \alpha_{(g*v,g*\mathfrak{a})}) : \alpha_{(v,\mathfrak{a})} \to \alpha_{(g*v,g*\mathfrak{a})}.$$

The basic result is the following. A proof is given in [24].

Theorem 4.17. *Given the stable object α in $\mathcal{R}$ with automorphism group G, a small disk structure* **D** *and a good deformation $v \to j(v)$ compatible with* **D***, there exists an open G-invariant neighborhood O of $(0,0) \in V \times \mathbb{B}^\alpha$ such that the following holds for a given gluing profile:*

(i) $\Psi : G \times O \to \mathcal{R}$ *is a fully faithful functor, injective on objects, and $\Psi(0,0) = \alpha$.*

(ii) *The map $|\Psi| : {}_G \backslash O \to |\mathcal{R}|$ induced on orbit spaces defines a homeomorphism onto an open neighborhood U of $|\alpha|$.*

(iii) *For every $(v, \mathfrak{a})$ the partial Kodaira–Spencer differentiable associated to $\alpha_{(v,\mathfrak{a})}$ is an isomorphism.*

(iv) *For every point $q \in O$, there exists an open neighborhood $U(q) \subset O$ with the property that every sequence $(q_k) \subset U(q)$ for which $|\Psi(q_k)|$ converges in $|\mathcal{R}|$ there exists a subsequence of (q_k) converging in $\mathrm{cl}_O(U(q))$.*

For the definition of the partial Kodaira–Spencer differentiable, see [30] or [24]. The above construction is possible for every given gluing profile φ.

Definition 4.18. *Assume a gluing profile φ is fixed. A functor $\Psi : G \times O \to \mathcal{R}$ constructed as described previously, associated to a stable α and a small disk structure $\mathbf{D}$ and satisfying the properties (i)–(iv) of Theorem 4.17 is called a good uniformizer associated to α.*

For the following discussion, we assume that a gluing profile φ has been fixed. Let us observe that the construction of $\Psi : G \times O \to \mathcal{R}$ is associated to a given object α in $\mathcal{R}$ and

$$\Psi(0,0) = \alpha \quad \text{and} \quad \Psi(g, (0,0)) = (\alpha, g, \alpha) : \alpha \to \alpha.$$

The construction requires the choice of a small disk structure $\mathbf{D}$, the good deformation $v \to j(v)$ and the open G-invariant subset O of $V \times \mathbb{B}^\alpha$. Here $\mathbb{B}^\alpha$ is a well-defined set associated to α and V is an open subset of the complex vector space $H^1(\alpha)$ associated to α. All in all, our possible choices involved in constructing Ψ for a fixed α constitute a set. Therefore, we can make the following definition.

Definition 4.19. *By $F(\alpha)$ we denote the set of good uniformizers associated to α.*

We shall elaborate next on the functorial properties of F. We can view the assignment

$$\alpha \to F(\alpha).$$

as a functor defined on $\mathcal{R}$ with values in the category of sets SET. We note that given an isomorphism $\Phi := (\alpha, \phi, \alpha') : \alpha \to \alpha'$, the choices made in the construction of $F(\alpha)$ correspond bijectively via ϕ to those made in the construction of $F(\alpha')$. In fact, there is a natural bijection $F(\Phi) : F(\alpha) \to F(\alpha')$. To be precise, recall that for $\alpha = (S, j, M, D)$ an element Ψ in the set $F(\alpha)$ is constructed by making the following choices:

(i) A small disk structure $\mathbf{D}$ for (S, j, M, D).

(ii) A deformation of j with specific properties $v \to j(v)$ on (S, M, D), i.e., that is constant on $\mathbf{D}$, has some symmetry properties and is defined on a G-invariant open neighborhood V of 0 in $H^1(\alpha)$.

(iii) From this data, we obtain the natural gluing parameters $\mathbb{B}^\alpha$ and can construct the family

$$\Psi : (v, \mathfrak{a}) \to (S_{\mathfrak{a}}, j(v)_{\mathfrak{a}}, M_{\mathfrak{a}}, D_{\mathfrak{a}}).$$

(iv) A G-invariant open neighborhood O of $(0, 0)$ in $V \times B^\alpha$ such that the restriction of the above family satisfies a list of properties defining an element in $F(\alpha)$; see Definition 4.18.

Hence the collection $F(\alpha)$ is obtained via a precise instruction that requires choices in its construction. When we consider isomorphic objects in $\mathcal{R}$, say α and α', there is a natural way of matching a specific choice for α with a specific choice for α'. This matching depends on the choice of an isomorphism and has functorial properties. With the morphism $\Phi := (\alpha, \phi, \alpha') : \alpha \to \alpha'$, we define the following data for α':

(i) $\mathbf{D}' := \phi_*(\mathbf{D}) := \{\phi(D_z) \mid z \in |D|\}$.

(ii) The biholomorphic map ϕ defines a linear isomorphism $\phi_* : H^1(\alpha) \to H^1(\alpha')$, and we take $V' = \phi_*(V)$, and a biholomorphic map $\phi_* : B^\alpha \to B^{\alpha'}$.

(iii) For j', we take the map $V' \ni v' \to j'(v') := T\phi \circ j(\phi_*^{-1} v') \circ T\phi^{-1}$.

(iv) $O' = \phi_*(O)$.

From the data $\mathbf{D}'$, $v' \to j'(v')$ and $\mathbb{B}^{\alpha'}$, we construct the element $\Psi' \in F(\alpha')$ corresponding to Ψ. This shows that a morphism $\Phi := (\alpha, \phi, \alpha') : \alpha \to \alpha'$ defines a bijection $F(\Phi) : F(\alpha) \to F(\alpha')$. Just saying that F is a functor from $\mathcal{R}$ into SET is again quite minimalistic, since we suppress the various maps ϕ_* we constructed above. Hence, underlying this bijection there is in our case even some fine structure, which we shall describe next.

Assume that the automorphism group of $\alpha = (S, j, M, D)$ is G. Then the elements of G have the form (α, g, α), where $g : \alpha \to \alpha$ is a biholomorphic map. For simplicity of notation, we identity $g = (\alpha, g, \alpha)$. Assume that we are given an isomorphism

$$\Phi = (\alpha, \phi, \alpha') : \alpha \to \alpha'.$$

Having the domains fixed (most importantly the almost complex structures), we identify $\phi = (\alpha, \phi, \alpha')$. Denoting by G' the automorphism group of α', and proceeding as in the case of G, the morphism Φ determines a group isomorphism

$$\gamma_\Phi : G \to G' : g \to \phi \circ g \circ \phi^{-1}.$$

Consider the (by the previous discussion) related elements $\Psi \in F(\alpha)$ and $\Psi' \in F(\alpha')$ given by

$$\Psi : G \ltimes O \to \mathcal{R} \quad \text{and} \quad \Psi' : G' \ltimes O \to \mathcal{R}.$$

Related means that $F(\Phi)(\Psi) = \Psi'$. The biholomorphic map ϕ defines an equivariant diffeomorphism $O \to O'$ that satisfies

$$\phi_*(g * q) = \gamma_\Phi(g) * \phi_*(q),$$

where $q = (v, a)$. This means in particular that ϕ_* defines a smooth equivalence between the translation groupoids, but also is a bijection on objects and morphisms:

$$f_\Phi : G \ltimes O \to G' \ltimes O' : (g, q) \to (\gamma_\Phi(g), \phi_*(q)).$$

From this, we see that we have the two functors

$$\Psi \quad \text{and} \quad \Psi' \circ f_\Phi,$$

both defined on $G \ltimes O$. On objects, with $(v', a') = f_\Phi(v, a)$, we have the biholomorphic map $\phi_a : \alpha_{(v,a)} \to \alpha'_{(v',a')}$, and hence the isomorphism

$$\Phi_{(v,a)} := (\alpha_{(v,a)}, \phi_a, \alpha'_{f_\Phi(v,a)}) : \alpha_{(v,a)} \to \alpha'_{f_\Phi(v,a)}.$$

Equivalently written, this precisely means, for an object (v, a) in $G \ltimes O$,

$$\Phi_{(v,a)} : \Psi(v, a) \to \Psi' \circ f_\Phi(v, a).$$

We note that $\Phi_{(0,0)} = \Phi$ and $f_\Phi(0,0) = (0,0)$. Assume that $(g, q) : q \to g * q$ is a morphism in $G \ltimes O$, with $(g', q') = f_\Phi(g, q)$ being the corresponding morphism in $G' \ltimes O'$. We obtain the following commutative diagram

$$
\begin{array}{ccc}
\Psi(q) & \xrightarrow{\ \Phi_q\ } & \Psi' \circ f_\Phi(q) \\
\Psi(g,q) \downarrow & & \downarrow \Psi' \circ f_\Phi(g,q) \\
\Psi(g * q) & \xrightarrow{\ \Phi_{g*q}\ } & \Psi' \circ f_\Phi(g * q)
\end{array}
$$

Hence the map

$$\Gamma : O \to \mathcal{R} : q \to \Gamma(q) = \Phi_q$$

is a natural transformation (in fact an equivalence)

$$\Psi \xrightarrow{\Gamma} \Psi' \circ f_\Phi.$$

This shows that there is more structure than just having a functor $F : \mathcal{R} \to \text{SET}$. Hence we have proved the following.

Theorem 4.20. *With the functor $F : \mathcal{R} \to \text{SET}$ as just described, given an isomorphism $\Phi : \alpha \to \alpha'$, there exists for every element $\Psi \in F(\alpha)$ with corresponding element $\Psi' = F(\Phi)(\Psi)$ a uniquely determined smooth equivalence of categories $f_\Phi : G \ltimes O \to G' \ltimes O'$ and a natural equivalence $\Gamma : \Psi \to \Psi' \circ f_\Phi$.*

At this point, the good uniformizers in $F(\alpha)$ and $F(\alpha')$ are mostly unrelated, with the exception of when the objects are isomorphic. In the following, in particular for the abstract theory, we shall always take the minimalistic viewpoint, even if in all applications there is additional structure along the lines just described.

We need some additional structure to relate $F(\alpha)$ and $F(\alpha')$ even if α and α' are not isomorphic. This needs some preparation.

Definition 4.21. *Given the functor $F : \mathcal{R} \to$ SET and objects α and α' in $\mathcal{R}$, we define for $\Psi \in F(\alpha)$ and $\Psi' \in F(\alpha')$ the set $\mathbf{M}(\Psi, \Psi')$, called the associated transition set between Ψ and Ψ', by*

$$\mathbf{M}(\Psi, \Psi') = \{(q, \Phi, q') \mid q \in O,\ q' \in O',\ \Phi : \Psi(q) \to \Psi'(q')\}.$$

The construction of the transition set comes with several structure maps.

Definition 4.22. *The target map $t : \mathbf{M}(\Psi, \Psi') \to O'$ and the source map $s : \mathbf{M}(\Psi, \Psi') \to O$ are given by*

$$t(q, \Phi, q') = q' \quad and \quad s(q, \Phi, q') = q.$$

The inversion map $\iota : \mathbf{M}(\Psi, \Psi') \to \mathbf{M}(\Psi', \Psi)$ is defined by $\iota(q, \Phi, q') = (q', \Phi^{-1}, q)$. The unit map $O \to \mathbf{M}(\Psi, \Psi)$ is given by $u(q) = (q, 1_{\Psi(q)}, q)$, and the multiplication map is defined as

$$m : \mathbf{M}(\Psi', \Psi'')_s \times_t \mathbf{M}(\Psi, \Psi') \to \mathbf{M}(\Psi, \Psi'') :$$

$$m((q', \Phi', q''), (q, \Phi, q')) = (q, \Phi' \circ \Phi, q'').$$

The following results hold for the logarithmic and the exponential gluing profiles. There are similar results for other, but not all, gluing profiles. The logarithmic case is a reformulation of the classical Deligne–Mumford theory. The case of the exponential gluing profile can be reduced to this classical case. In [24] we derive both cases using PDE methods, but also show that the case with exponential gluing profile can be reduced to the classical case.

Theorem 4.23. *Given the logarithmic or the exponential gluing profile, the transition sets $\mathbf{M}(\Psi, \Psi')$ carry a natural metrizable topology and in addition naturally oriented smooth manifold structures, so that all the structure maps are smooth, and s and t are orientation preserving local diffeomorphisms. In the case of the logarithmic gluing profile, the smooth manifold structure underlies a complex manifold structure and all structure maps are holomorphic. The object space of $G \ltimes O$ is oriented by the orientation coming from the fact that O is an open subset of a complex vector space. The orientations of the manifolds $\mathbf{M}(\Psi, \Psi')$ are determined by the fact that the source and target maps are orientation-preserving.*

In particular, at this point, we have, after picking the exponential gluing profile φ, a natural construction $(F, \mathbf{M})$ that associates to every object a set $F(\alpha)$ of good uniformizers Ψ at α, and to a transition set $\mathbf{M}(\Psi, \Psi')$ an oriented manifold structure.

Remark 4.24. Let us just note that with Ψ' corresponding to Ψ if $\Phi : \alpha \to \alpha'$ is given, the natural transformation $\Gamma : \Psi \to \Psi' \circ f_\Phi$ defines a diffeomorphism

$$G \ltimes O \to \mathbf{M}(\Psi, \Psi') : (g, q) \to (q, \Gamma(g * q) \circ \Psi(g, q), f_\Phi(g * q)).$$

Here is one thing we can do with this data. Take a family $(\Psi_\lambda)_{\lambda \in \Lambda}$ (Λ a set) of good uniformizers such that

$$|\mathcal{R}| = \bigcup_{\lambda \in \Lambda} |\Psi_\lambda(O_\lambda)|.$$

Then we define an oriented smooth manifold X by

$$X = \coprod_{\lambda \in \Lambda} O_\lambda,$$

and a smooth oriented manifold $\mathbf{X}$ by

$$\mathbf{X} = \coprod_{(\lambda, \lambda') \in \Lambda \times \Lambda} \mathbf{M}(\Psi_\lambda, \Psi_{\lambda'}).$$

We may view X as the collection of objects in a small category and $\mathbf{X}$ as the set of morphisms. More precisely, if $q_\lambda \in O_\lambda$ and $q'_{\lambda'} \in O_{\lambda'}$, then the morphisms $q_\lambda \to q'_{\lambda'}$ are precisely all elements $\Phi \in \mathbf{M}(\Psi_\lambda, \Psi_{\lambda'})$ with $s(\Phi) = q_\lambda$ and $t(\Phi) = q'_{\lambda'}$.

Using property (iv), it follows immediately that $\mathcal{X} = (X, \mathbf{X})$ is an étale proper Lie groupoid—see [48] for a short description of the theory. More comprehensive treatments are given in [1], and see [19–21] for the beginnings of this theory. Using the Ψ_λ, we can construct a functor

$$\beta : \mathcal{X} \to \mathcal{R},$$

which is an equivalence of categories. Namely, we map the object $q_\lambda \in O_\lambda \subset X$ to $\Psi_\lambda(q_\lambda)$, so that on objects

$$\beta : X \to \text{obj}(\mathcal{R}) : \beta(q_\lambda) = \Psi_\lambda(q_\lambda),$$

and on morphisms

$$\beta : \mathbf{X} \to \text{mor}(\mathcal{R}) : \beta(q_\lambda, \phi, q'_{\lambda'}) = (\Psi_\lambda(q_\lambda), \phi, \Psi_{\lambda'}(q'_{\lambda'})).$$

Hence we might view $\mathcal{X}$ as a smooth (up to equivalence) model for $\mathcal{R}$. Of course, making different choices, we obtain $\beta' : \mathcal{X}' \to \mathcal{R}$ having the same properties. Taking the union of the choices, we obtain $\beta'' : \mathcal{X}'' \to \mathcal{R}$ and smooth equivalences of étale proper

Lie groupoids via the inclusions $\mathcal{X} \to \mathcal{X}''$ and $\mathcal{X}' \to \mathcal{X}''$. Taking the weak fibered product associated to the diagram

$$\mathcal{X} \to \mathcal{X}'' \leftarrow \mathcal{X}'$$

and denoted by $\mathcal{X} \times_{\mathcal{X}''} \mathcal{X}'$, the projections onto the factors are equivalences of étale proper Lie groupoids. In other words, $\mathcal{X}$ and $\mathcal{X}'$ are Morita-equivalent. Moreover, $\beta :$ $\mathcal{X} \to \mathcal{R}$ and $\beta' : \mathcal{X}' \to \mathcal{R}$ are naturally equivalent via the data from $\mathcal{X}''$. In summary, we have constructed up to Morita equivalence smooth models for $\mathcal{R}$. Each pair $(\mathcal{X}, |\beta|)$ gives an orbifold structure on $|\mathcal{R}|$ and these orbifold structures are equivalent. The whole collection turns $|\mathcal{R}|$ into a smooth oriented orbifold, with structure depending on the gluing profile φ. If the process is carried out with the logarithmic gluing profile, we again obtain a natural orbifold $|\mathcal{R}|$, which this time, however, is holomorphic. The underlying smooth structures for the two constructions mentioned are not the same, but they should be diffeomorphic. The latter is definitely true over the un-noded part. The reader is referred to $[1, 48, 49]$ for more reading on étale proper Lie groupoids.

4.2.2 The category of stable maps

The construction carried out in this subsection is formally very similar to the one we carried out for $\mathcal{R}$. Now the category $\mathcal{R}$ is replaced by a different category $\mathcal{S}$, the category of stable maps. Again $|\mathcal{S}|$ has a natural metrizable topology and we shall construct a pair $(F, \mathbf{M})$ similarly as before. However, the local models are infinite-dimensional and none of the ingredients have any chance to be classically smooth. In fact, the maps that occur are generally nowhere-differentiable in any classical sense. Nevertheless, as it will turn out, there is a smoothness concept, called sc-smoothness, such that the constructions viewed within an associated differential geometry are smooth.

Let (Q, ω) be a compact symplectic manifold without boundary. We consider maps defined on Riemann surfaces with image in (Q, ω) having various regularity properties. We shall write

$$u : \mathcal{O}(S, x) \to Q$$

for a mapping germ defined on a Riemann surface S near x.

Definition 4.25. *Let $m \geq 2$ be an integer and $\varepsilon > 0$. We say a germ of a continuous map $u : \mathcal{O}(S, x) \to Q$ is of class (m, ε) at the point x if for a smooth chart $\phi : U(u(0)) \to \mathbb{R}^{2n}$ mapping $u(0)$ to 0 and holomorphic polar coordinates $\sigma : [0, \infty) \times S^1 \to S \setminus \{x\}$ around x, the map*

$$v(s, t) = \phi \circ u \circ \sigma(s, t),$$

defined for s large, has partial derivatives up to order m, which, weighted by $e^{\varepsilon s}$, belong to $L^2([s_0, \infty) \times S^1, \mathbb{R}^{2n})$ if s_0 is sufficiently large. We say that the germ is of class m around a point $z \in S$ provided u is of class H^m_{loc} near z.

We observe that the above definition does not depend on the choices involved, like charts and holomorphic polar coordinates. We consider now tuples $\alpha = (S, j, M, D, u) = (\alpha^*, u)$, where $\alpha^* = (S, j, M, D)$ is a noded Riemann surface with ordered marked points M and nodal pairs D, and $u : S \to W$ is a continuous map having some additional regularity properties. We do not assume α^* to be stable, i.e., to be an object in $\mathcal{R}$ after forgetting the order of M.

Definition 4.26. *A noded Riemann surface with ordered marked points is a tuple $\alpha^* = (S, j, M, D)$, where (S, j) is a closed Riemann surface, $M \subset S$ a finite collection of ordered marked points, and D is a finite collection of unordered pairs $\{x, y\}$ of points in S, called nodal pairs, such that $x \neq y$ and two pairs that intersect are identical. The union of all $\{x, y\}$, denoted by $|D|$, is disjoint from M. We call D the set of nodal pairs and $|D|$ the set of nodal points.*

As in the definition of the objects in $\mathcal{R}$, the Riemann surface S might consist of different connected components C. We call C a domain component of S. The special points on C are, as before, the points in $C \cap (M \cup |D|)$. We say that (S, j, M, D) is connected, provided the topological space $\bar{S}$ obtained by identifying $x \equiv y$ in the nodal pairs $\{x, y\} \in D$ is connected. With our terminology, it is possible that $\alpha^* = (S, j, M, D)$ is connected, but on the other hand S may have several connected components, i.e., its domain components.

Next we describe the tuples α in more detail.

Definition 4.27. *We say that $\alpha = (S, j, M, D, u)$ is a stable map of class (m, δ) provided the following hold, where $m \geq 2$ and $\delta > 0$:*

 (i) *The underlying topological space obtained by identifying the two points in any nodal pair is connected.*

 (ii) *The map u is of class (m, δ) around the points in $|D|$ and of class m around all other points in S.*

 (iii) *$u(x) = u(y)$ for every nodal pair $\{x, y\} \in D$.*

 (iv) *If a domain component C of S has genus g_C and n_C special points such that $2g_C + n_C \leq 2$, then $\int_C u^*\omega > 0$. Otherwise, we assume that $\int_C u^*\omega \geq 0$. This is called the stability condition.*

Next we introduce the category of stable maps of class $(3, \delta_0)$, where δ_0 is a fixed number in $(0, 2\pi)$.

Definition 4.28. *Fix a $\delta_0 \in (0, 2\pi)$. The category of stable maps (of class $(3, \delta_0)$), denoted by*

$$\mathcal{S}^{3,\delta_0}(Q, \omega),$$

has as objects the tuples α of class $(3, \delta_0)$ and as morphisms

$$\Phi := (\alpha, \phi, \alpha') : \alpha \to \alpha',$$

where ϕ is a biholomorphic map $(S, j, M, D) \rightarrow (S', j', M', D')$ satisfying $u' \circ \phi = u$.

Remarks 4.29.

(i) The category $\mathcal{S}^{3,\delta_0}(Q, \omega)$ has several interesting features. All the morphisms are isomorphisms, and between two objects the number of morphisms is finite. This is a consequence of the stability condition.

(ii) If we identify isomorphic objects in $\mathcal{S}^{3,\delta_0}(Q, \omega)$, we obtain the orbit class $|\mathcal{S}^{3,\delta_0}(Q, \omega)|$, which is easily verified to be a set. An element in the orbit set is written as $|\alpha|$.

(iii) The choice of $\delta_0 \in (0, 2\pi)$ is dictated by the fact that in any analytical treatment of stable maps (Fredholm theory), one has to derive elliptic estimates near the nodes that involve self-adjoint operators that have spectrum $2\pi \mathbb{Z}$. So $(0, 2\pi)$ has to be understood as a spectral gap. This requirement is important for the Fredholm theory.

The first important fact is that the orbit set $|\mathcal{S}^{3,\delta_0}(Q, \omega)|$ carries a natural topology. More precisely, we have the following.

Theorem 4.30. *Given $\delta_0 \in (0, 2\pi)$, the orbit space $|\mathcal{S}^{3,\delta_0}(Q, \omega)|$ carries a natural Hausdorff, second-countable, regular and hence metrizable topology.*

We refer the reader to [12] or [30] for the complete construction. However, we shall give some ideas later on. The key is a recipe to construct a basis for a topology. This recipe involves choices. Nevertheless, the resulting topology is independent of these, and this is what we mean by it being natural. However, this is only the beginning and there is another natural construction. What we shall see next will remind the reader of the type of construction that already occurred in the treatment of $\mathcal{R}$. In the case of $\mathcal{R}$, we saw uniformizers defined on translation groupoids, where object and morphism sets carried natural smooth manifold structures. In the relevant constructions for $\mathcal{S}^{3,\delta_0}(Q, \omega)$, our translation groupoids will have much less structure (for the moment); namely, they are just metrizable topological spaces. The reason is that all the maps that occur have no chance of being classically smooth and the natural local models seem very often far away from open subsets of Banach spaces.

Recall that given a metrizable space O with the action of a group G by continuous maps, we can construct the metrizable translation groupoid $G \ltimes O$. This is a small metrizable category with the objects being the elements in O and the morphisms being the tuples (g, q), where $q \in O$ and $g \in G$. Here the source of (g, q) is q and the target is $g * q$, i.e.,

$$(g, q) : q \rightarrow g * q.$$

A metrizable category is a small category, where object and morphism sets are metrizable spaces and all structure maps are continuous, namely, associating to a morphism ϕ its source and target, the inversion map $\phi \rightarrow \phi^{-1}$ and the unit map $q \rightarrow 1_q$, as well as the multiplications map $m(\phi, \psi) = \phi \circ \psi$ defined on the appropriate subspace of the product of two copies of the morphism space.

Definition 4.31. *Let α be an object in $\mathcal{S}^{3,\delta_0}(Q, \omega)$ with automorphism group G. A good uniformizer around α is a functor $\Psi : G \ltimes O \to \mathcal{S}^{3,\delta_0}(Q, \omega)$ having the following properties:*

(i) *Ψ is fully faithful, injective on objects, and there exists an object $q_0 \in O$ with $\Psi(q_0) = \alpha$.*

(ii) *Passing to orbit spaces, $|\Psi| :\ _G\backslash O \to |\mathcal{S}^{3,\delta_0}(Q, \omega)|$ is a homeomorphism onto an open neighborhood of $|\alpha|$ in $|\mathcal{S}^{3,\delta_0}(Q, \omega)|$.*

Given two good uniformizers Ψ around α and Ψ' around α', we can define the set $\mathbf{M}(\Psi, \Psi')$ consisting of all tuples (q, Φ, q'), where $q \in O$, $q' \in O'$ and $\Phi : \Psi(\alpha) \to \Psi'(\alpha')$ is a morphism in $\mathcal{S}^{3,\delta_0}(Q, \omega)$. Associated to $\mathbf{M}(\Psi, \Psi')$ we have the source map

$$s : \mathbf{M}(\Psi, \Psi') \to O : (q, \Phi, q') \to q$$

and the target map

$$t : \mathbf{M}(\Psi, \Psi') \to O' : (q, \Phi, q') \to q'.$$

In addition, there is the inversion map

$$\iota : \mathbf{M}(\Psi, \Psi') \to \mathbf{M}(\Psi', \Psi) : (q, \Phi, q') \to (q', \Phi^{-1}, q),$$

the unit map

$$u : O \to \mathbf{M}(\Psi, \Psi) : q \to (q, 1_{\Psi(q)}, q)$$

and, given a third uniformizer Ψ'', the multiplication map

$$m : \mathbf{M}(\Psi', \Psi'')_s \times_t \mathbf{M}(\Psi, \Psi') \to \mathbf{M}(\Psi, \Psi'') :$$

$$m((q', \Phi', q''), (q, \Phi, q')) = (q, \Phi' \circ \Phi, q'').$$

The maps s, t, ι, u and m are called the structure maps. The next natural construction is summarized by the following theorem, where, as before, SET is the category of sets.

Theorem 4.32. *There exists a natural functor $F : \mathcal{S}^{3,\delta_0}(Q, \omega) \to$ SET that associates to an object α a set $F(\alpha)$ of good uniformizers around α. Moreover, F comes with a natural construction, which associates to every choice $\Psi \in F(\alpha)$ and $\Psi' \in F(\alpha')$ a metrizable topology on $\mathbf{M}(\Psi, \Psi')$ such that the source and target maps are local homeomorphisms and all structure maps are continuous.*

Remark 4.33. One can be more explicit about the correspondence of constructions for α and the constructions for α', when we are given a morphism $\Phi : \alpha \to \alpha'$. This is in the spirit of the discussion for $\mathcal{R}$.

There are many consequences of this result. For example, we can pick a family $(\Psi_\lambda)_{\lambda \in \Lambda}$ (where Λ is a set) of good uniformizers such that

$$|S^{3,\delta_0}(Q,\omega)| = \bigcup_{\lambda \in \Lambda} |\Psi(O_\lambda)|.$$

Then we can define the objects of a category X as the disjoint union

$$X = \coprod_{\lambda \in \Lambda} O_\lambda$$

and the morphism set $\mathbf{X}$ by

$$\mathbf{X} = \coprod_{(\lambda,\lambda') \in \Lambda \times \Lambda} \mathbf{M}(\Psi_\lambda, \Psi_{\lambda'}).$$

We note that X and $\mathbf{X}$ are both metrizable, and the Ψ_λ can be used to define a functor β from $\mathcal{X} = (X, \mathbf{X})$ to $S^{3,\delta_0}(Q,\omega)$ by

$$X \to \mathrm{obj}(S^{3,\delta_0}(Q,\omega)) : O_\lambda \ni q \to \Psi_\lambda(q)$$

and

$$\mathbf{X} \to \mathrm{mor}(S^{3,\delta_0}(Q,\omega)) : \quad \mathbf{M}(\Psi_\lambda, \Psi_{\lambda'}) \ni (q, \Phi, q') \to \Phi.$$

Lemma 4.34. *The functor β is an equivalence of categories.*

In other words, we can build a topological version of $S^{3,\delta_0}(Q,\omega)$ up to equivalence. Of course, the construction of $\mathcal{X}$ involves a choice of a family (Ψ_λ). As in the $\mathcal{R}$ case, if we make a different choice, we obtain

$$\beta' : \mathcal{X} \to S^{3,\delta_0}(Q,\omega)$$

and, taking the disjoint union of the choices, $\beta'' : \mathcal{X}'' \to S^{3,\delta_0}(Q,\omega)$. The inclusions $\mathcal{X} \to \mathcal{X}''$ and $\mathcal{X}' \to \mathcal{X}''$ are local homeomorphisms on the object and morphism spaces, and equivalences of categories.

Starting from this situation and quite similar to the $\mathcal{R}$ case, one can define a notion of equivalence of two pairs $(\mathcal{X}, \beta)$ and $(\mathcal{X}', \beta')$, so that we might say that $S^{3,\delta_0}(Q,\omega)$ up to Morita equivalence has a unique topological (metrizable) model. As it turns out, the construction of F in Theorem 4.32 has many more properties. Namely, the domains $G \ltimes O$ of the functors Ψ_λ are in some sense smooth spaces—not in the usual way, but in a generalized differential geometry. This will also be explained later on. At the end of the day, there is a differential geometric version of Theorem 4.32, which one might view as the construction of a smooth structure on the category $S^{3,\delta_0}(Q,\omega)$, and which allows the construction of small smooth versions of our category up to Morita equivalence.

4.2.3 A bundle and the CR-functor

Assume next that a compatible almost complex structure J has been fixed for (Q,ω). We consider tuples

$$\widehat{\alpha} = (\alpha, \xi) = (S, j, M, D, u, \xi),$$

where α is an object in $S^{3,\delta_0}(Q,\omega)$ and $\xi(z) : T_z S \to T_{u(z)}Q$ is complex antilinear for the given structures j and J. Further, we assume that

$$z \to \xi(z)$$

has Sobolev regularity H^2 away from the nodal points. At the nodal points, we assume it to be of class $(2,\delta_0)$. To make this precise, pick a nodal point x and take positive holomorphic polar coordinates around x, say $\sigma(s,t)$ with $x = \lim_{s\to\infty} \sigma(s,t)$. Then take a smooth chart ϕ around $u(x)$ with $\phi(u(x)) = 0$. Finally, consider the principal part

$$(s,t) \to pr_2 \circ T\phi(u(\sigma(s,t))) \circ \xi(\sigma(s,t)) \left(\frac{\partial\sigma(s,t)}{\partial s} \right),$$

which we assume for large s_0 to be in $H^{2,\delta_0}([s_0,\infty) \times S^1, \mathbb{R}^{2n})$. The definition of the decay property does not depend on the choice of σ and ϕ. Denote by BAN_G the category whose objects are the Banach spaces and whose morphisms are topological linear isomorphisms. The above discussion gives us a functor

$$\mu : S^{3,\delta_0}(Q,\omega) \to \mathrm{BAN}_G.$$

We associate to an object α the Hilbert space of all ξ of class $(2,\delta_0)$ as described above. To a morphism $\Phi = (\alpha, \phi, \alpha')$ we associate the topological linear isomorphism

$$\mu(\alpha) \to \mu(\alpha') : \xi \to \xi \circ T\phi^{-1}.$$

We can now define a new category $\mathcal{E}^{2,\delta_0}(Q,\omega,J)$ whose objects are the tuples $\widehat{\alpha}$ of the form

$$\widehat{\alpha} = (\alpha, \xi),$$

where α is an object in $S^{3,\delta_0}(Q,\omega)$ and $\xi \in \mu(\alpha)$ is a complex antilinear TQ-valued $(0,1)$-form along u of class $(2,\delta_0)$. A morphism

$$\widehat{\Phi} = (\widehat{\alpha}, \Phi, \widehat{\alpha}') : \widehat{\alpha} \to \widehat{\alpha}'$$

is a tuple with $\Phi = (\alpha, \phi, \alpha') : \alpha \to \alpha'$ being a morphisms in $\mathcal{S}^{3,\delta_0}(Q, \omega)$ such that, in addition,

$$\mu(\Phi)(\xi) = \xi'.$$

There is the projection functor

$$P : \mathcal{E}^{2,\delta_0}(Q, \omega, J) \to \mathcal{S}^{3,\delta_0}(Q, \omega),$$

which on objects maps (α, ξ) to α and on morphisms maps $((\alpha, \xi), \Phi, (\alpha', \xi'))$ to Φ. On the object level, the preimage under P of an object α is a Hilbert space consisting of the elements (α, ξ). Given $\Phi : \alpha \to \alpha'$, the morphism Φ is lifted to a bounded linear isomorphism

$$\Phi : P^{-1}(\alpha) \to P^{-1}(\alpha') : (\alpha, \xi) \to (\alpha', \xi \circ T\phi^{-1}),$$

where the map $\xi \to \xi \circ T\phi^{-1}$ is linear. We may view the two categories fibering over each other via $P : \mathcal{E}^{2,\delta_0}(Q, \omega, J) \to \mathcal{S}^{2,\delta_0}(Q, \omega)$ as a "bundle." Then we have the Cauchy–Riemann section functor

$$\bar{\partial}_J(\alpha) = \left(\alpha, \tfrac{1}{2}[Tu + J(u) \circ Tu \circ j]\right),$$

where $\alpha = (S, j, M, D, u)$. From the results in [30], we can deduce the following theorem.

Theorem 4.35. *The orbit space $|\mathcal{E}^{2,\delta_0}(Q, \omega, J)|$ carries a natural second-countable met-rizable topology. The induced maps on the orbit spaces $|P|$ and $|\bar{\partial}_J|$ are continuous. The topology on the orbit space of the full subcategory associated to the objects α with $\bar{\partial}_J(\alpha) = 0$ has compact connected components and coincides with the (union of all) Gromov-compactified moduli space of J-pseudoholomorphic curves.*

The previous discussion about the construction F can be extended to cover P. Given an object α, the notion of a good uniformizer generalizes by replacing O by a continuous surjective map $p : W \to O$ between metrizable spaces, where the fibers are Hilbert spaces (or more generally Banach spaces), and where in addition we have a continuous G-action, linear between the fibers. Then the good bundle uniformizers are given by commutative diagrams

$$
\begin{array}{ccc}
G \ltimes W & \xrightarrow{\bar{\Psi}} & \mathcal{E}^{2,\delta_0}(Q, \omega, J) \\
\downarrow & & \downarrow \\
G \ltimes O & \xrightarrow{\Psi} & \mathcal{S}^{3,\delta_0}(Q, \omega)
\end{array}
$$

with the obvious properties. The bottom Ψ is as described before and $\bar{\Psi}$ is a linear bounded isomorphism between the fibers. Again, as it will turn out, these constructions will fit into a scheme of generalized differential geometry, and the local representative of $\bar{\partial}_J$, a G-equivariant section of $W \to O$, will turn out to be a nonlinear Fredholm section in the extended framework. We refer the reader to $[13, 30]$ for the constructions related to stable maps and to $[31, 32]$ for the abstract theory. In the above extension, we have a functorial construction $\bar{F} : S^{3,\delta_0}(Q, \omega) \to \mathrm{SET}$ covering F in the following way. For every object α, the set $\bar{F}(\alpha)$ consists of strong bundle uniformizers $\bar{\Psi}$, which, together with the underlying $\Psi \in F(\alpha)$, fit into the above commutative diagram. Moreover, it comes with an associated construction of a metrizable topology on the transition set $\mathbf{M}(\bar{\Psi}, \bar{\Psi}')$. There is a projection

$$\mathbf{p} : \mathbf{M}(\bar{\Psi}, \bar{\Psi}') \to \mathbf{M}(\Psi, \Psi') : (k, \widehat{\Phi}, k') \to (o, \Phi, o')$$

whose fibers are Hilbert spaces. Moreover, there exists the following commutative diagram involving source and target maps:

$$
\begin{array}{ccccc}
K & \xleftarrow{\ s\ } & \mathbf{M}(\bar{\Psi}, \bar{\Psi}') & \xrightarrow{\ t\ } & K \\
{\scriptstyle p}\downarrow & & {\scriptstyle p}\downarrow & & {\scriptstyle p'}\downarrow \\
O & \xleftarrow{\ s\ } & \mathbf{M}(\Psi, \Psi') & \xrightarrow{\ t\ } & O'
\end{array}
$$

We leave the execution of the idea to the reader. The above follows from the results in $[30]$.

4.2.4 A basic construction

The following discussion is carried out in detail in $[33]$; see also $[30]$. Consider the Hilbert space E consisting of pair of maps (u^+, u^-)

$$u^{\pm} : \mathbb{R}^{\pm} \times S^1 \to \mathbb{R}^N,$$

where for each pair there exists a constant $c \in \mathbb{R}^N$ such that $u^{\pm} - c$ has partial derivatives up to order 3 that, weighted by $e^{\delta_0|s|}$, belong to $L^2(\mathbb{R}^{\pm} \times S^1)$. The constant c is called the common asymptotic limit of (u^+, u^-).

Given a complex number $|a| < \frac{1}{2}$, we write $a = |a| e^{-2\pi i \theta}$. If $a \neq 0$, we define $R = \varphi(|a|)$, where φ is the exponential gluing profile, which we recall, is defined by

$$\varphi : (0, 1] \to [0, \infty) : \varphi(r) = e^{1/r} - e.$$

We construct for $0 < |a| < \frac{1}{2}$ the finite cylinder Z_a by identifying $(s, t) \in [0, R] \times S^1$ with $(s', t') \in [-R, 0] \times S^1$ via

$$s = s' + R \quad \text{and} \quad t = t' + \theta.$$

Denote by $[s, t]$ the equivalence class of the point $(s, t) \in [0, R] \times S^1$ and by $[s', t']'$ the equivalence class of $(s', t') \in [-R, 0] \times S^1$. Then

$$[s, t] = [s - R, t - \theta]'$$

and $Z_a \to [0, R] \times S^1$ and $Z_a \to [-R, 0] \times S^1$ defined by $[s, t] \to (s, t)$ and $[s', t']' \to (s', t')$, respectively, are holomorphic coordinates, where the targets are equipped with the standard complex structures.

If $a = 0$, we define $Z_0 = \mathbb{R}^+ \times S^1 \bigsqcup \mathbb{R}^- \times S^1$. Pick a smooth map $\beta : \mathbb{R} \to [0, 1]$ satisfying $\beta(s) = 1$ for $s \leq -1$, $\beta'(s) < 0$ for $s \in (-1, 1)$ and $\beta(s) + \beta(-s) = 1$ for all $s \in \mathbb{R}$.

Definition 4.36. *Given $(u^+, u^-) \in E$ and $|a| < \frac{1}{2}$, the glued map*

$$\oplus_a(u^+, u^-) : Z_a \to \mathbb{R}^N,$$

is defined for $a = 0$ by $\oplus_0(u^+, u^-) = (u^+, u^-)$ and for $a \neq 0$ by

$$\oplus_a(u^+, u^-)([s, t]) = \beta(s - \tfrac{1}{2}R)u^+(s, t) + \left(1 - \beta(s - \tfrac{1}{2}R)\right) u^-(s - R, t - \theta).$$

For fixed $a \neq 0$, many different (u^+, u^-) will produce the same glued map. We can remove the ambiguity by introducing the anti-glued map $\ominus_a(u^+, u^-)$. It is defined on C_a given by $C_0 = \varnothing$ and for $0 < |a| < \frac{1}{2}$ by gluing $\mathbb{R}^- \times S^1$ and $\mathbb{R}^+ \times S^1$ along Z_a. Note that Z_a can be identified with an obvious subset of both half-cylinders. Namely, C_a is given by

$$C_a = \left((\mathbb{R}^- \times S^1) \bigsqcup (\mathbb{R} \times S^1)\right) / \sim,$$

where $(s', t') \equiv (s, t)$ provided $(s', t') \in [-R, 0] \times S^1$, $(s, t) \in \mathbb{R}^+ \times S^1$, and $s = s' + R$ and $t = t' + \theta$. We shall write $[s, t]$ or equivalently $[s', t']'$ for the points in the part being identified. Note that we have for $0 < |a| < \frac{1}{2}$ a natural embedding

$$Z_a \to C_a : [s, t] \to [s, t].$$

For $a \neq 0$, the cylinder C_a has a natural complex manifold structure, for which the maps on $Z_a \subset C_a$ defined by

$$[s, t] \to (s, t) \in \mathbb{R} \times S^1$$

and

$$[s', t']' \to (s', t') \in \mathbb{R} \times S^1$$

have natural extensions to biholomorphic maps $C_a \to \mathbb{R} \times S^1$. The extensions will be written as above. The transition map

$$\mathbb{R} \times S^1 \to \mathbb{R} \times S^1 : (s, t) \to (s', t')$$

is given by $(s,t) \to (s-R, t-\theta)$.

Definition 4.37. *For* $(u^+, u^-) \in E$, *the anti-glued map* $\ominus_a(u^+, u^-) : C_a \to \mathbb{R}^N$ *is defined for* $a = 0$ *by* $\ominus_0(u^+, u^-) = 0$, *i.e., the only map* $\varnothing \to \mathbb{R}^N$, *and otherwise by*

$$\ominus_a(u^+, u^-)([s,t]) = \left(\beta(s - \tfrac{1}{2}R) - 1\right) \left(u^+(s,t) - av_a(u^+, u^-)\right)$$
$$+ \beta(s - \tfrac{1}{2}R)\left(u^-(s-R, t-\theta) - av_a(u^+, u^-)\right).$$

Here

$$av_a(u^+, u^-) = \tfrac{1}{2}\int_{S^1} \left(u^+(\tfrac{1}{2}R, t) + u^-(-\tfrac{1}{2}R, t)\right) dt.$$

If we start with $(u^+, u^-) \in E$ and pick any $0 < |a| < \tfrac{1}{2}$, the total gluing map

$$\square_a(u^+, u^-) = (\oplus_a(u^+, u^-), \ominus_a(u^+, u^-)),$$

defines a bounded linear isomorphism from E to $H^3(Z_a, \mathbb{R}^N) \oplus H^{3,\delta_0}_c(C_a, \mathbb{R}^N)$. Here $H^{3,\delta_0}_c(C_a, \mathbb{R}^N)$ denotes the Hilbert space of maps $u : C_a \to \mathbb{R}^N$ that are of class H^3_{loc}, so that in addition there exists a constant $c \in \mathbb{R}^N$ depending on u for which $u \pm c$ has partial derivatives up to order 3 that, weighted by $e^{\delta_0|s|}$, belong to $L^2([s_0, \infty) \times S^1)$ or $L^2(-\infty, -s_0] \times S^1)$, respectively, for a suitably large s_0. In particular, we obtain for $a \neq 0$ a linear topological isomorphism

$$\ker(\ominus_a) \to H^3(Z_a, \mathbb{R}^N) : (u^-, u^+) \to \oplus_a(u^+, u^-).$$

Since E decomposes as the topological direct sum

$$E = \ker(\ominus_a) \oplus \ker(\oplus_a),$$

there exists an associated family $a \to \pi_a$ of bounded projection operators $\pi_a : E \to E$ that project onto $\ker(\ominus_a)$ along $\ker(\oplus_a)$.

Proposition 4.38. *For every* $|a| < \tfrac{1}{2}$, *the linear operator* $\pi_a : E \to E$ *is a bounded projection. However, the map*

$$\{a \in \mathbb{C} \,|\, |a| < \tfrac{1}{2}\} \to L(E) : a \to \pi_a,$$

where the space of bounded operators $L(E)$ *is equipped with the operator norm, is nowhere continuous. The map*

$$\{a \in \mathbb{C} \,|\, |a| < \tfrac{1}{2}\} \times E \to E : (a, (u^+, u^-)) \to \pi_a(u^+, u^-)$$

is continuous.

The map $r : \{|a| < \tfrac{1}{2}\} \times E \to \{|a| < \tfrac{1}{2}\} \times E$ defined by

$$r(a, (u^+, u^-)) = (a, \pi_a(u^+, u^-))$$

is continuous and satisfies $r \circ r = r$. It also preserves the fibers of $\{|a| < \frac{1}{2}\} \times E \to$ $\{|a| < \frac{1}{2}\}$. The image of r is then a fiber-wise retract. Define $\ker(\ominus.)$ as the subset of $\{a \in \mathbb{C} \mid |a| < \frac{1}{2}\} \times E$ consisting of all tuples $(a, (u^+, u^-))$ with $\ominus_a(u^+, u^-) = 0$. Then we have a continuous projection

$$\ker(\ominus.) \to \{a \in \mathbb{C} \mid |a| < \frac{1}{2}\},$$

where the fibers are sc-Hilbert spaces. Let us define the set $\bar{X}^{3,\delta_0}(\mathbb{R}^N)$ by

$$\bar{X}^{3,\delta_0}(\mathbb{R}^N) := (\{0\} \times E) \bigcup \left(\bigcup_{0 < |a| < \frac{1}{2}} (\{0\} \times H^3(Z_a, \mathbb{R}^N)) \right).$$

We obtain a fiber-preserving, fiber-wise linear, bijection

$$\ker(\ominus.) \xrightarrow{\quad \oplus. \quad} \bar{X}^{3,\delta_0}(\mathbb{R}^N)$$

$$\downarrow \qquad\qquad\qquad \downarrow$$

$$\{|a| < \tfrac{1}{2}\} =\!=\!=\!= \{|a| < \tfrac{1}{2}\}$$

We equip the right-hand side with the topology $\mathcal{T}$ that makes $\oplus.$ a homeomorphism. We observe that $\ker(\ominus.)$ depends on the choice of the cut-off function β, whereas the right-hand side does not depend on any choice. So one might ask if $\mathcal{T}$ depends on the choice of β.

Proposition 4.39. *The definition of the topology $\mathcal{T}$ does not depend on the choice of the cut-off function β. Moreover, $\mathcal{T}$ is a metrizable topology.*

As a consequence of this proposition, $\bar{X}^{3,\delta_0}(\mathbb{R}^N)$ is in a natural way a metrizable topological space such that the projection

$$p : \bar{X}^{3,\delta_0}(\mathbb{R}^N) \to \{|a| < \tfrac{1}{2}\} : [u : Z_a \to \mathbb{R}^N] \to a$$

is continuous. We can carry out this construction for every $\mathbb{R}^N$. Interestingly, a smooth map $f : \mathbb{R}^N \to \mathbb{R}^M$ induces a continuous map

$$\bar{X}^{3,\delta_0}(\mathbb{R}^N) \to \bar{X}^{3,\delta_0}(\mathbb{R}^M) : u \to f \circ u.$$

In summary, this means that the following assertion holds true.

Proposition 4.40. *The construction $\bar{X}^{3,\delta_0}$ defines a natural functor from the category of Euclidean spaces with objects $\mathbb{R}^N$ and smooth maps between them into the category of metrizable topological spaces.*

If Q is a smooth connected manifold without boundary, then we can take for a sufficiently large N a smooth proper embedding $j : Q \to \mathbb{R}^N$ and define $\bar{X}^{3,\delta_0}(Q)$ to consist of all continuous maps $u : Z_a \to Q$ such that $j \circ u \in X^{3,\delta_0}(\mathbb{R}^N)$, and equip it with the topology making the map $u \to j \circ u$ a homeomorphism onto its image. The resulting topological space $\bar{X}^{3,\delta_0}(Q)$ does not depend on the choice of the embedding. Indeed, if $k : Q \to \mathbb{R}^M$ is another proper embedding, then $k \circ j^{-1}$ and $j \circ k^{-1}$ are restrictions of globally defined smooth maps, which implies the conclusion. If Q does not have boundary but is not connected, we can apply the above to every component. Consequently, we have the following result.

Proposition 4.41. *The functor $\bar{X}^{3,\delta_0}$, defined for the spaces $\mathbb{R}^N$ and the smooth maps between them, has a natural extension to the category of smooth manifolds without boundaries and the smooth maps between them.*

Given a finite-dimensional smooth manifold L and a smooth map

$$\Lambda : L \to \{a \in \mathbb{C} \,|\, |a| < \tfrac{1}{2}, \}$$

we can consider the pullback $\Lambda^* X^{3,\delta_0}(\mathbb{R}^N) \subset \Lambda \times X^{3,\delta_0}(\mathbb{R}^N)$ defined by the diagram

$$X^{3,\delta_0}(\mathbb{R}^N)$$
$$\downarrow$$
$$L \xrightarrow{\ \Lambda\ } \{|a| < \tfrac{1}{2}\}$$

The next result is more complicated to prove. Denote by j the standard almost complex structure on Z_a. We assume that for $|a| < \tfrac{1}{2}$ we are given a family

$$a \to j(a)$$

of smooth almost complex structures on Z_a. The $j(a)$ live on different domains and we require a certain form of smooth dependence on a. The first requirement is the following, where φ denotes the exponential gluing profile.

(i) There exist $\varepsilon \in (0, \tfrac{1}{2})$ and $H > 0$ such that

$$j(a) = j \text{ on } \{[s,t] \in Z_a \,|\, s \in [H, \varphi(|a|) - H]\}$$

provided $0 < |a| < \varepsilon$, and if $a = 0$, then it holds that $j(0)|(\mathbb{R}^{\pm} \times S^1)$ equals j on $[H, \infty) \times S^1$ and $(-\infty, -H] \times S^1$. Here we assume that $\varphi(\varepsilon) > 2H$.

In other words, if the neck gets longer, the structure will be j in the necks, but can be different near the boundaries. In the next step, we shall have to formulate the smoothness requirements near the boundaries. Consider for $|a| < \varepsilon$ the maps

$$[0, \tfrac{3}{2}H] \times S^1 \to Z_a : (s,t) \to [s,t]$$

and

$$[-\tfrac{3}{2}H, 0] \times S^1 \to Z_a : (s', t') \to [s', t']',$$

which we can use to pull back $j(a)$ to obtain the families $j^+(a)$ and $j^-(a)$, respectively, defined for $|a| < \varepsilon$.

(ii) For ε as in (i), the maps $a \to j^{\pm}(a)$ and $|a| < \varepsilon$ are smooth.

Next we need to say something about the smoothness for $|a| < \tfrac{1}{2}$. If $\tfrac{1}{2}\varepsilon < |a| < \tfrac{1}{2}$, we can identify $[0, 1] \times S^1$ with Z_a via the map $(s, t) \to [\varphi(|a|)s, t]$ and, via pullback, obtain the family $j^+(a)$, $\tfrac{1}{2}\varepsilon < |a| < \tfrac{1}{2}$ on $[0, 1] \times S^1$. Similarly, we can identify $[-1, 0] \times S^1$ with Z_a via the map $(s', t') \to [-\varphi(|a_0|)s', t']'$ and obtain the pullback family $a \to j^-(a)$ on $[-1, 0] \times S^1$. We require the following.

(iii) $a \to j^+(a)$ and $a \to j^-(a)$ for $\tfrac{1}{2}\varepsilon < |a| < \tfrac{1}{2}$ are smooth families.

Definition 4.42. *A family $a \to j(a)$ that associates to $|a| < \tfrac{1}{2}$ a smooth almost complex structure on Z_a and having the properties (i)–(iii) as described above will be called a* good family *(of smooth almost complex structures).*

There are many equivalent formulations for defining a good family. The important result is concerned with the continuity of a certain class of maps.

Theorem 4.43. *Assume j and k are two good families as described above and*

$$\{|a| < \tfrac{1}{2}\} \to \{|a| < \tfrac{1}{2}\} : a \to b(a)$$

is a smooth map such that there exists a core-smooth family (defined below)

$$a \to [\phi_a : (Z_a, j(a)) \to (Z_{b(a)}, k(b(a)))]$$

of biholomorphic maps. Then the map

$$b^*\bar{X}^{3,\delta_0}(Q) \to \bar{X}^{3,\delta_0}(Q) : (a, u) \to u \circ \phi_a$$

is continuous.

The family $a \to \phi_a$ being core-smooth just requires this map and its inverse to be smooth near the boundaries. This is well defined, using the coordinates $[s, t]$ and $[s', t']'$. By unique continuation, ϕ_a is entirely determined by knowing a, $b(a)$ and its values near the boundaries. So the fact that $\phi_a : Z_a \to Z_{b(a)}$ is biholomorphic also determines the behavior inside, which, of course, is crucial for the validity of the theorem.

Remark 4.44. The proof of this result requires some work. The above map will in fact be sc-smooth once we have introduced this concept. Proofs of sc-smooth versions of Theorem 4.43 can be found In [30, 33]. We note that maps of the type occurring in

Theorem 4.43 will usually not be classically differentiable unless they are constant. In fact, the expression $u \circ \phi_a$, when differentiated with respect to a, would lose a derivative of u.

The previous discussion can be extended. In a first step, rather than considering the half-cylinders $\mathbb{R}^{\pm} \times S^1$, we consider a noded disk pair $(D_x \cup D_y, \{x, y\})$. From all possible decorations $[\widehat{x}, \widehat{y}]$ of $\{x, y\}$, we obtain the natural gluing parameters $r[\widehat{x}, \widehat{y}]$ with $r \in [0, \frac{1}{2})$ and define the set of gluing parameters associated to $\{x, y\} \in D$ as described before by

$$\mathbb{B}^{\{x,y\}}\left(\tfrac{1}{2}\right) = \{r[\widehat{x}, \widehat{y}] \mid r \in [0, \tfrac{1}{2}), \ [\widehat{x}, \widehat{y}] \text{ decorated nodal pair}\}.$$

If we pick $\widehat{x}$ and $\widehat{y}$, so that $\{\widehat{x}, \widehat{y}\}$ is a representative of $[\widehat{x}, \widehat{y}]$, there are (unique) associated biholomorphic maps $h_x : (D_x, x) \to (\mathbf{D}, 0)$ and similarly h_y so that $Th_x(x)\widehat{x} = \mathbb{R}$ and $Th_y(y)\widehat{y} = \mathbb{R}$. With this data, we first obtain the map

$$\phi : \mathbb{B}^{\{x,y\}}\left(\tfrac{1}{2}\right) \to \{z \in \mathbb{C} \mid |z| < \tfrac{1}{2}\} : r[\widehat{x}, \theta\widehat{y}] \to r\theta,$$

which, by the definition of the complex structure on $\mathbb{B}^{\{x,y\}}(\frac{1}{2})$, is biholomorphic. If $\mathfrak{a} = [\widehat{x}, \theta\widehat{y}]$ and h_x and h_y are as described above, then h_x and $\theta^{-1}h_y$ is the uniquely determined choice of maps $(D_x, x) \to (\mathbb{D}, 0)$ and $(D_y, y) \to (\mathbb{D}, 0)$, with $Th_x(x)\widehat{x} = \mathbb{R}$ and $T(\theta^{-1}h_y)(y)(\theta\widehat{y}) = \mathbb{R}$. Hence the identification $z \equiv z'$ with $h_x(z)h_y(z') = \theta e^{-2\pi R}$ defines $Z_{r[\widehat{x}, \theta\widehat{y}]}$, where $R = \varphi(r)$. With h_x and h_y fixed, let us define biholomorphic maps

$$D_x \setminus \{x\} \to [0, \infty) \times S^1 : z \to (s, t), \qquad \text{where} \quad e^{-2\pi(s+it)} = h_x(z),$$

and

$$D_y \setminus \{y\} \to (-\infty, 0] \times S^1 : z' \to (s', t'), \qquad \text{where} \quad e^{2\pi(s'+it')} = h_y(z').$$

We would like to find the complex number $a = a(\mathfrak{a})$ such that we obtain a natural induced biholomorphic map from $Z_{\mathfrak{a}}$ to Z_a. Clearly, the identification $z \equiv z'$ associated to $h_x(z)(\theta^{-1}h_y(z')) = e^{-2\pi R}$, which defines $Z_{\mathfrak{a}}$ if $\mathfrak{a} = r[\widehat{x}, \theta\widehat{y}]$, is equivalent to the identification $(s, t) \equiv (s', t')$ given by

$$e^{-2\pi(s+it)}e^{2\pi(s'+it')} = e^{-2\pi i\vartheta}e^{-2\pi R},$$

where we write θ as $e^{-2\pi i\vartheta}$. Hence

$$s = s' + R \quad \text{and} \quad t = t' + \vartheta.$$

This is the identification defining Z_a. Hence, with $a(r[\widehat{x}, \theta\widehat{y}]) = r\theta$, we obtain an induced biholomorphic map

$$Z_{\mathfrak{a}} \to Z_{a(\mathfrak{a})} : [z = e^{-2\pi(s+it)}] \to [s, t].$$

This family of maps allows us to use the definition of $\bar{X}^{3,\delta_0}(\mathbb{R}^N)$ to define a set of maps on the $\mathcal{Z}_\mathfrak{a}$ resulting in a set $\bar{X}_{\mathcal{D}}^{3,\delta_0}(\mathbb{R}^N)$ with

$$\mathcal{D} = (D_x \cup D_y, \{x, y\}).$$

More precisely,

$$\bar{X}_{\mathcal{D}}^{3,\delta_0}(\mathbb{R}^N) = H_c^{3,\delta_0}(\mathcal{D}, \mathbb{R}^N) \coprod \left[\coprod_{\mathfrak{a} \in \mathbb{B}^{\{x,y\}}(\frac{1}{2}) \mid |0<\mathfrak{a}|<\frac{1}{2}\}} H^3(Z_\mathfrak{a}, \mathbb{R}^N) \right].$$

The elements in $H_c^{3,\delta_0}(\mathcal{D}, \mathbb{R}^N)$ are pairs (u^+, u^-) of continuous maps defined on D_x and D_y, respectively, such that $u^+(x) = u^-(y)$ and if σ_x are positive polar coordinates on D_x centered at x, and σ_y negative ones on D_y centered at y, then the pair $(u^+ \circ \sigma_x, u^- \circ \sigma_y)$ belongs to E. The previous construction might a priori depend on the choices involved. However, making different initial choices of identifying $\bar{X}_{\mathcal{D}}^{3,\delta_0}(\mathbb{R}^N)$ with E, the associated transition families induce continuous maps as a consequence of Theorem 4.43. This is a natural construction and the space itself is homeomorphic to the image of a continuous retraction. Moreover, it follows immediately from the definition that for a smooth map $f : \mathbb{R}^N \to \mathbb{R}^M$, the map $w \to f \circ w$ is continuous. From this, we can deduce that the functor $\bar{X}_{\mathcal{D}}^{3,\delta_0}$ has a natural extension to all smooth manifolds Q without boundary.

At this point, we have a well-defined functorial construction that associates to a nodal disk pair $\mathcal{D} = (D_x \cup D_y, \{x, y\})$ and smooth manifold without boundary Q a metrizable topological space $\bar{X}_{\mathcal{D}}^{3,\delta_0}(Q)$ and to a smooth map $f : Q \to Q'$ a continuous map

$$\bar{X}_{\mathcal{D}}^{3,\delta_0}(Q) \to \bar{X}_{\mathcal{D}}^{3,\delta_0}(Q') : w \to f \circ w.$$

We can go one step further. This step is natural as well. Consider a noded Riemann surface (S, j, D). We assume that S has no boundary and is compact. We do not make any stability assumption. Suppose that we are given a finite group G acting by biholomorphic maps on (S, j, D) and $\mathbf{D}$ is a small disk structure invariant under G. Then we can take the natural gluing parameters and consider the family $\mathfrak{a} \to (S_\mathfrak{a}, j_\mathfrak{a}, D_\mathfrak{a})$. We have dealt with problems arising near the nodes and the obtained glued necks. Away from these regions, we are just dealing with "static" domains and Sobolev class H^3 functions on them. We obtain the following result from the previous discussion.

Theorem 4.45. *Given a noded closed Riemann surface (S, j, D) with an action by biholomorphic maps by a finite group and a G-invariant small disk structure, there exists a functorial construction of a metrizable topological space $\bar{X}_{(S,j,D),\mathbf{D}}^{3,\delta_0}(Q)$ for every smooth manifold Q without boundary. Moreover, G acts continuously on these spaces. Associated to a smooth map $f : Q \to Q'$ there is a continuous equivariant map*

$$\bar{X}_{(S,j,D),\mathbf{D}}^{3,\delta_0}(Q) \to \bar{X}_{(S,j,D),\mathbf{D}}^{3,\delta_0}(Q') : w \to f \circ w.$$

There are some additional useful constructions that will be used later. Assume that $\Xi \subset S$ is a finite G-invariant subset in the complement of the disks coming from the small disk structure. Suppose that we have fixed for every G-orbit $[z]$ of a point $z \in \Xi$ a codimension-2 submanifold $H_{[z]}$ in Q. Denote the collection of all these constraints by $\mathcal{H}$. Define a subset $\bar{X}^{3,\delta_0}_{(S,j,D),\mathbf{D},\mathcal{H}}(Q)$ of $\bar{X}^{3,\delta_0}_{(S,j,D),\mathbf{D}}(Q)$ to consist of all u intersecting $H_{[z]}$ at z transversally for all $z \in \Xi$. If we have a map u in $\bar{X}^{3,\delta_0}_{(S,j,D),\mathbf{D},\mathcal{H}}(Q)$, then we find a small neighborhood around every $z \in \Xi$ and an open neighborhood U around u such that for every $z \in \Xi$ there is continuous map $U \ni w \to z_w$, uniquely determined by the requirement that w intersects $H_{[z]}$ transversally at z_w. Denote by Ξ' a deformation of Ξ. We can construct a continuous family of self-maps $\phi_{\Xi',a} : (S_a, D_a) \to (S_a, D_a)$, which in the necks are the identity and which move Ξ' back to Ξ and are supported in a small neighborhood around $\Xi..$ Then the map defined for $w \in U$ given by

$$w \to w \circ \phi^{-1}_{\Xi_w,a},$$

where Ξ_w is the deformation associated to w, i.e., the collection of all z_w, defines a continuous retraction onto some open neighborhood $O = O(w)$ such that $r(O) = O \cap \bar{X}^{3,\delta_0}_{(S,j,D),\mathbf{D},\mathcal{H}}(Q)$. In other words, the subset $\bar{X}^{3,\delta_0}_{(S,j,D),\mathbf{D},\mathcal{H}}(Q)$ of $\bar{X}^{3,\delta_0}_{(S,j,D),\mathbf{D}}(Q)$ is locally a retract given by a retraction of a particular form. Since the ambient space was modeled locally on retracts, the same is true for this subset.

Remark 4.46. None of the constructions that we have carried out has a chance to be classically smooth. However, as we shall see later, they are in fact smooth constructions in the sc-smooth world, i.e., they are differential geometric constructions in an extended differential geometry as described in [31, 32].

We complete the discussion by adding some words about the construction of certain bundles over $\bar{X}^{3,\delta_0}_{(S,j,D),\mathbf{D},\mathcal{H}}(Q)$. First of all, we would like to change the almost complex structure on the (S_a, D_a) by deforming the j_a. Let us assume for the moment that we are given a smooth family

$$j : V \ni v \to j(v)$$

satisfying $j(0) = j$ and $j(v) = j$ on the disks of the small disk structure. Here V is an open subset of some finite-dimensional vector space. In general, we would have an action of G on these spaces, but at the moment this is not relevant for the point we would like to make. Then $V \times \bar{X}^{3,\delta_0}_{(S,j,D),\mathbf{D},\mathcal{H}}(Q)$ is a metrizable topological space and we can view a point (v, w) in this space as a map defined on $(S_a, j(v)_a, D_a)$, i.e., defined on a deformation of (S_a, j_a, D_a), the space on which w was originally defined. We consider next tuples (v, u, ξ), where $z \to \xi(z)$ is of class H^{2,δ_0} and for fixed $z \in S_a$ it holds that

$$\xi(z) : (T_z S_a, j(v)_a) \to (T_{u(z)} Q, J)$$

is complex antilinear. We denote by $\bar{E}^{2,\delta_0}_{(S,j,D),\mathbf{D},\mathcal{H},j}(Q, \omega, J)$ the collection of all these tuples. We have a natural projection

$$p : \bar{E}^{2,\delta_0}_{(S,j,D),\mathbf{D},\mathcal{H},j}(Q,\omega,J) \rightarrow V \times \bar{X}^{3,\delta_0}_{(S,j,D),\mathbf{D},\mathcal{H}}(Q) : (v,u,\xi) \rightarrow (v,u).$$

A similar construction involving retractions allows us to equip the above with a natural metrizable topology and the local models for these spaces are $K \rightarrow O$, where O is a retraction in a Hilbert space, and similarly for K in a product of Hilbert spaces, where in this case the retraction is linear on fibers, with respect to a projection of the product onto the first factor. We refer the reader to [30, 33] for details.

4.2.5 A good uniformizer

In the following discussion, we shall need the version of Deligne–Mumford theory of stable Riemann surfaces using the exponential gluing profile φ. The discussion is in spirit completely parallel to that for the category $\mathcal{R}$. However, the difference is that all maps occurring in the discussion of $\mathcal{R}$ are either smooth or holomorphic (depending on the gluing profile), but in the stable-map discussion they are only continuous.

We shall now describe the construction of the functors Ψ that were introduced in Section 4.2.2. Fix an object α in $\mathcal{S}^{3,\delta_0}(Q,\omega)$ with automorphism group G and write $\alpha = (S,j,M,D,u)$. As a consequence of the stability condition, we obtain the following.

Lemma 4.47. *The group G is a finite group.*

We need the notion of a domain stabilization.

Definition 4.48. *A domain stabilization Ξ for α consists of a finite subset $\Xi \subset S \setminus (M \cup |D|)$ such that the following holds:*

    (i) (S,j,M^*,D) with M^* being the unordered set $M \cup \Xi$ is a stable, perhaps noded, Riemann surface.

    (ii) Ξ is an invariant set under the action of G.

    (iii) If $z, z' \in \Xi$ and $u(z) = u(z')$, then there exists a $g \in G$ with $g(z) = z'$.

    (iv) For every $z \in \Xi$ the map $Tu(z) : T_z S \rightarrow T_{u(z)} Q$ is injective and pulls back ω to a non-degenerate two-form on $T_z S$ defining the same orientation as j.

One easily proves that domain stabilizations always exist as a consequence of the stability assumption.

Lemma 4.49. *Every object α in $\mathcal{S}^{3,\delta_0}(Q,\omega)$ has a domain stabilization.*

After fixing a domain stabilization, we have two stable objects, namely, the stable map α in $\mathrm{obj}(\mathcal{S}^{3,\delta_0}(Q))$ and the stable noded Riemann surface with unordered marked points $\alpha^* = (S,j,M \cup \Xi, D)$, which is an object in $\mathcal{R}$. The automorphism group G of α is contained in the automorphism group G^* of α^*, and both are finite groups.

We continue with α^* and pick a small disk structure $\mathbf{D}$ that is invariant under G^*. Then we fix a good deformation $v \rightarrow j(v)$, with $v \in V$, where $V \subset H^1(\alpha^*)$ is an open G^*-invariant subset. We can construct the functor associated to

$$(v,a) \rightarrow (S_a, j(v)_a, M^*_a, D_a),$$

where $M^* = M \cup \Xi$. Taking a sufficiently small G^*-invariant open neighborhood O of $(0,0)$ in $H^1(\alpha^*) \times \mathbb{B}^{\alpha^*}$, we obtain a good uniformizer

$$\Psi^* : G^* \ltimes O \to \mathcal{R}.$$

Denote the G-orbit for $z \in \Xi$ by $[z]$. It follows that if $[z] \neq [z']$, then $u(z) \neq u(z')$. Of course, u takes the same value on all the elements of $[z]$. For every $[z]$, we fix a smooth oriented submanifold $H_{[z]}$ in Q of codimension 2 so that $u(z) \in H_{[z]}$ and the oriented range$(Tu(z)) \oplus H_{[z]}$ equals the oriented $T_{u(z)}Q$. Then u intersects $H_{[z]}$ transversally at z. We assume that $H_{[z]}$ does not have a boundary. It is even fine to assume that it is diffeomorphic to an open ball. Denote the whole collection $(H_{[z]})$ by $\mathcal{H}$.

Recall our topological space $\bar{X}^{3,\delta_0}_{(S,j,D),D}(Q)$, which contains $u : (S,D) \to Q$. From the Sobolev embedding theorem, we know that $H^3_{\text{loc}} \to C^1_{\text{loc}}$. As a consequence, the subset $\bar{X}^{3,\delta_0}_{(S,j,D),D,\mathcal{H}}(Q)$, consisting of all $w : (S_a, D_a) \to Q$ that at $z \in \Xi$ intersect $H_{[z]}$ transversally, is well defined. We also recall that if $w \in \bar{X}^{3,\delta_0}_{(S,j,D),D}(Q)$ is close enough to u, then there is a unique intersection point z_w near z such that w intersects $H_{[z]}$ transversally. Moreover, $w \to z_w$ is continuous. One can use this to prove the following previously mentioned result.

Proposition 4.50. *For every map w in $\bar{X}^{3,\delta_0}_{(S,j,D),D,\mathcal{H}}(Q)$, there exists an open neighborhood $U(w)$ in $\bar{X}^{3,\delta_0}_{(S,j,D),D}(Q)$ and a continuous map $r : U(w) \to U(w)$ such that $r(U(w)) = U(w) \cap \bar{X}^{3,\delta_0}_{(S,j,D),D,\mathcal{H}}(Q)$.*

In other words, the subspace associated to the constraints is a local continuous retract. This part is very important and ties in with the Deligne–Mumford theory. With our original object being the stable map $\alpha = (S, j, M, D, u)$, it holds that $(S, D, u) \in \bar{X}^{3,\delta_0}_{(S,j,D),D,\mathcal{H}}(Q)$. We also have the object α^* in $\mathcal{R}$, and with the small disk structure we obtained the good uniformizing family

$$(a, v) \to (S_a, j(v)_a, (M \cup \Xi)_a, D_a)$$

that was used to construct the good uniformizer

$$\Psi^* : G^* \ltimes O^* \to \mathcal{R}.$$

Next we form the topological product space $V \times \bar{X}^{3,\delta_0}_{(S,j,D),D,\mathcal{H}}(Q)$ and bring the two separate discussions together. We define for its elements the map

$$(v, (S_a, D_a, w)) \to (S_a, j(v)_a, M_a, D_a, w).$$

The group G acts on the topological space $V \times \bar{X}^{3,\delta_0}_{(S,j,D),D,\mathcal{H}}(Q)$, and the above map defines a functor

$$\Psi : G \ltimes (V \times \bar{X}^{3,\delta_0}_{(S,j,D),D,\mathcal{H}}(Q)) \to \mathcal{S}^{3,\delta_0}(Q).$$

The fundamental observation is the following, which follows from the results in [30].

Theorem 4.51. *For a suitable G-invariant open neighborhood O of*

$$(0, (S, D, u)) \in V \times \bar{X}^{3,\delta_0}_{(S,j,D),\mathbf{D},\mathcal{H}}(Q),$$

the following properties hold:

(i) $\Psi : G \times O \to S^{3,\delta_0}(Q)$ *is full and faithful, and injective on objects.*

(ii) *The map induced on orbit spaces is a homeomorphism onto an open neighborhood of* $|\alpha|$.

(iii) *If $(v, (S_\mathfrak{a}, D_\mathfrak{a}, w)) \in O$, then $(\mathfrak{a}, v) \in O^*$, where we recall the good uniformizer for $\mathcal{R}$ denoted by $\Psi^* : G^* \times O^* \to \mathcal{R}$.*

We note that the construction of Ψ requires several choices. However, starting with an object α, we obtain a set worth $F(\alpha)$ of functors having the properties stated in the theorem. Hence again we obtain a functorial construction

$$F : S^{3,\delta_0}(Q, \omega) \to \text{SET}.$$

Remark 4.52. Moreover, if $\Phi : \alpha \to \alpha'$ is an isomorphism, the possible choices made for α can be bijectively identified with choices for α'. This means that F is, in fact, a functor $S^{3,\delta_0}(Q, \omega) \to \text{SET}$. Further, there is a precise geometric relationship between Ψ and $\Psi' = F(\Phi)(\Psi)$, similar to the $\mathcal{R}$-case.

Assume that we are given two such functors Ψ and Ψ' associated to stable maps α and α' and suppose there exist q_0 and q'_0 for which there exists an isomorphism

$$\Phi_0 : \Psi(q_0) \to \Psi'(q'_0).$$

Then Φ_0 has an underlying biholomorphic map ϕ_0 and, as we shall see, taking a variation q' of q'_0, there is a uniquely determined core-continuous (see [30]) germ $q' \to \phi_{q'}$, and a germ of homeomorphism $q \to q(q')$ such that $\Phi_{q'} := (\Psi(q(q')), \phi_{q'}, \Psi'(q'))$ satisfies

$$\Phi_{q'} : \Psi(q(q')) \to \Psi'(q').$$

One can use this fact to construct a topology on the transition sets $\mathbf{M}(\Psi, \Psi')$. For this topology, the source and target maps will be local homeomorphisms and all structure maps will be continuous. One can find the details in [30], where this program is carried out in the sc-smooth case, which we shall discuss later on. The current topological discussion is easier.

In summary, there is a natural construction $(F, \mathbf{M})$ for $S^{3,\delta_0}(Q, \omega)$. Here F associates to an object α a set $F(\alpha)$ of good uniformizers defined on metrizable translation groupoids, and $\mathbf{M}$ associates to the transition sets $\mathbf{M}(\Psi, \Psi')$ metrizable topologies, having the property that the structural maps become continuous and the source and target maps even local homeomorphisms.

The procedure that we have just outlined can be extended to deal with

$$P : \mathcal{E}^{2,\delta_0}(Q, \omega, J) \to \mathcal{S}^{3,\delta_0}(Q, \omega).$$

In this case, the construction $(F, \mathbf{M})$ can be extended to a construction $(\bar{F}, \bar{\mathbf{M}})$ covering the construction $(F, \mathbf{M})$, where $\bar{F} : \mathcal{S}^{3,\delta_0}(Q, \omega) \to$ SET, but associates to an object α a good bundle uniformizer

$$
\begin{array}{ccc}
G \ltimes K & \xrightarrow{\bar{\Psi}} & \mathcal{E}^{2,\delta_0}(Q, \omega, J) \\
p \downarrow & & p \downarrow \\
G \ltimes O & \xrightarrow{\Psi} & \mathcal{E}^{3,\delta_0}(Q, \omega)
\end{array}
$$

It requires variations of the previous constructions—the necessary details can be found in [30].

One might raise the question, given the naturality of the construction, whether there is perhaps more structure to be found—in particular, since perturbations of the pseudo-holomorphic part of $\mathcal{S}^{3,\delta_0}(Q, \omega)$ for a given compatible almost complex structure are used to define Gromov–Witten invariants. The answer is a resounding "yes." In fact, all the constructions that have occurred are smooth constructions in an extension of differential geometry that relies on a more flexible notion of differentiability in Banach spaces. In this differential geometry, there is a much larger library of smooth finite-dimensional or infinite-dimensional local models for smooth spaces. These local models can have locally varying dimensions but still have tangent spaces. Moreover, there is a nonlinear Fredholm theory with the usual expected properties. We shall describe this in the next section.

4.3 Smoothness

It is an amazing fact that the construction $(F, \mathbf{M})$ is not just topological, but in fact a smooth construction within a suitable framework of smoothness, which is quite different from the classical one. It will be described next. We keep in mind the occurrence of continuous retractions when constructing the domains of the good uniformizers.

4.3.1 sc-structures and sc-smooth maps

Assume we are given two Banach spaces E and F for which we have as vector spaces a continuous inclusion $E \subset F$. In interpolation theory [59], general methods are developed to construct Banach spaces that interpolate between E and F. We take the concept of a scale (with suitable properties) from interpolation theory, but give it a new interpretation as a generalization of a smooth structure. This study was initiated in [25–27], and leads to a quite unexpectedly rich theory. In [31, 32], we streamlined the presentation and added many further developments, which have not been published before.

Definition 4.53. *Let E be a Banach space. An sc-smooth structure (or sc-structure for short) for E consists of a nested sequence of Banach spaces $E_0 \supset E_1 \supset E_2 \supset \cdots$ with $E_0 = E$ so that*

1. *The inclusion $E_{i+1} \to E_i$ is a compact operator.*
2. *$E_\infty = \bigcap_i E_i$ is dense in every E_m.*

Example 4.54. *A typical example is $E = L^2(\mathbb{R})$ with the sc-structure given by $E_m := H^{m,\delta_m}(\mathbb{R})$, where $H^{m,\delta_m}(\mathbb{R})$ is the Sobolev space of functions in L^2 such that the derivatives up to order m weighted by $e^{\delta_m|s|}$ belong to L^2. Here δ_m is a strictly increasing sequence starting with $\delta_0 = 0$.*

If E and F are sc-Banach spaces, then $E \oplus F$ has a natural sc-structure given by

$$(E \oplus F)_m = E_m \oplus F_m.$$

Every finite-dimensional vector space has a unique sc-structure, namely, the constant one, where $E_i = E$. If E is infinite-dimensional, then the constant sequence violates Definition 4.53(1). We continue with some considerations about linear sc-theory.

Definition 4.55. *Let E be an sc-Banach space and $F \subset E$ a linear subspace. We call F an sc-subspace provided the filtration $F_i = F \cap E_i$ turns F into an sc-Banach space. If $F \subset E$ is an sc-Banach space, then we say that it has an sc-complement, provided there exists an sc-subspace G such $F_i \oplus G_i = E_i$ as topological linear sum for all i.*

Let us note that a finite-dimensional subspace F of E has an sc-complement if and only if $F \subset E_\infty$; see [25].

The linear operators of interest are those linear operators $T : E \to F$, that map E_m into F_m for all m, such that $T : E_m \to F_m$ is a bounded linear operator. We call T an sc-operator. An sc-isomorphism $T : E \to F$ is a bijective sc-operator whose inverse is also an sc-operator. Of particular interest are the linear sc-Fredholm operators.

Definition 4.56. *A sc-operator $T : E \to F$ is said to be sc-Fredholm provided there exist sc-splittings $E = K \oplus X$ and $F = Y \oplus C$ such that C and K are smooth and finite-dimensional, $Y = T(X)$, and $T : X \to Y$ defines a linear sc-isomorphism.*

We note that the above implies that $E_m = X_m \oplus K$ and $F_m = T(X_m) \oplus C$ for all m. The Fredholm index is by definition

$$\mathrm{Ind}(T) = \dim(K) - \dim(C).$$

Let us also observe that for every m, we have a linear Fredholm operator (in the classical sense) $T : E_m \to F_m$, which in particular has the same index and identical kernel.

Next we begin with the preparations to introduce the notion of an sc-smooth map. If $U \subset E$ is an open subset, then we can define an sc-structure for U by the nested sequence $(U_i)_{i=0}^\infty$ given by $U_i = E_i \cap U$. We note that $U_\infty = \bigcap U_i$ is dense in every U_m. Considering U with its sc-structure, we see that $U_{i_0} \subset E_{i_0}$ also admits an sc-structure defined by

$$(U_{i_0})_m := U_{i_0+m}.$$

We write U^{i_0} for U_{i_0} equipped with this sc-structure. Given two such sc-spaces U and V, we write $U \oplus V$ for $U \times V$ equipped with the obvious sc-structure. Now we can give the rigorous definition of the tangent TU of an open subset U in an sc-Banach space E.

Definition 4.57. *The tangent TU of an open subset $U \subset E$ of the sc-Banach space E is defined by $TU = U^1 \oplus E$.*

We note that

$$(TU)_i = U_{1+i} \oplus E_i.$$

Continuing Example 4.54, we have

$$TL^2(\mathbb{R}) = H^{1,\delta_1}(\mathbb{R}) \oplus L^2(\mathbb{R}) \quad \text{and} \quad (TL^2(\mathbb{R}))_i = H^{i+1,\delta_{i+1}}(\mathbb{R}) \oplus H^{i,\delta_i}(\mathbb{R}).$$

Definition 4.58. *Given two open subsets U and V in sc-Banach spaces, a map $f : U \to V$ is said to be of class sc^0 provided that for every m, the map f maps U_m into V_m and the map $f : U_m \to V_m$ is continuous.*

The following example takes a little bit of work.

Example 4.59. *Take $L^2(\mathbb{R})$ with the previously defined sc-structure and define*

$$\Phi : \mathbb{R} \oplus L^2(\mathbb{R}) \to L^2(\mathbb{R}) : (t, u) \to \Phi(t, u),$$

where $\Phi(t, u)(s) = u(s + t)$. Then Φ is sc^0.

Next we define the notion of an sc^1-map.

Definition 4.60. *Let $U \subset E$ and $V \subset F$ be open subsets in sc-Banach spaces. An sc^0-map $f : U \to V$ is said to be sc^1 provided that for every $x \in U_1$, there exists a continuous linear operator $Df(x) : E_0 \to F_0$ such that the following holds:*

1. *For $h \in E_1$ with $x + h \in U$, we have*

$$\lim_{\|h\|_1 \to 0} \frac{1}{\|h\|_1} \|f(x + h) - f(x) - Df(x)h\|_0 = 0.$$

2. *The map Tf defined by $Tf(x, h) = (f(x), Df(x)h)$ for $(x, h) \in TU$ defines an sc^0-map $Tf : TU \to TV$.*

Inductively, we can define the notion of a sc^k-map. A map is sc^∞ provided it is sc^k for all k. The following result shows that the chain rule holds. This is quite unexpected, since this fact looks not compatible with Definition 4.60(1). However, one is saved by the compactness of the inclusions stipulated by our definition of sc-structure.

Theorem 4.61 (Chain rule). *Assume that U, V and W are open subsets in sc-Banach spaces and that $f : U \to V$ and $g : V \to W$ are sc^1-maps. Then $g \circ f$ is sc^1 and $T(g \circ f) = (Tg) \circ (Tf)$. The same holds for sc^k and sc-smooth maps.*

Example 4.62. *One can show [33] that the map Φ from Example 4.59 is sc-smooth. Classically, it is nowhere differentiable.*

4.3.2 sc-smooth spaces and M-polyfolds

Now we are in a position to introduce new local models for smooth spaces. The interesting thing about sc-smoothness is the fact that there are many smooth retractions with complicated images, so that one obtains a large "library" of smooth local models for spaces. This library is large enough to describe problems occurring when studying the nonlinear Cauchy–Riemann operators in symplectic geometry, which shows analytical limiting behavior allowing for bubbling-off and similar analytical phenomena.

Definition 4.63. *Let $U \subset E$ be an open subset of the sc-Banach space E. A map $r : U \to U$ is called an sc^{∞}-retraction provided that it is sc-smooth and $r \circ r = r$.*

The chain rule implies that for an sc^{∞}-retraction r, its tangent map Tr is again an sc^{∞}-retraction. We call the image $O = r(U)$ of an sc^{∞}-retraction $r : U \to U$ an sc^{∞}-retract. The crucial definition is the following.

Definition 4.64. *A local sc-model is a pair (O, E), where E is an sc-Banach space and $O \subset E$ an sc^{∞}-retract given as the image of an sc-smooth retraction $r : U \to U$ defined on an open subset U of E.*

The following lemma is easily established.

Lemma 4.65. *Assume that (O, E) is a local sc-model and r and s are sc-smooth retractions defined on open subsets U and V of E, respectively, having O as the image. Then $Tr(TU) = Ts(TV)$.*

In view of this lemma, we can define the tangent of a local sc-model, which again is a local sc-model, as follows.

Definition 4.66. *The tangent of the local sc-model (O, E) is defined by*

$$T(O, E) := (TO, TE),$$

where $TO = Tr(TU)$ for any sc-smooth retraction $r : U \to U$ having O as the image, where U is open in E.

Remark 4.67. *Let us observe that if (O, E) is a local sc-model and O' an open subset of O, then (O', E) is again a local sc-model. Indeed, if $r : U \to U$ is an sc-smooth retraction with $O = r(U)$, then define $U' = r^{-1}(O')$, and $r' = r|U' : U' \to U'$ is an sc-smooth retraction with image O'.*

A map $f : O \to O'$ between two local sc-models is sc-smooth (or sc^k) provided that $f \circ r : U \to E'$ is sc-smooth (or sc^k). One easily verifies that the definition does not depend on the choice of r. We can define the tangent map $Tf : TO \to TO'$ by

$$Tf := T(f \circ r)|Tr(TU).$$

As it turns out, this is well defined and does not depend on the choice of r as long as it is compatible with (O, E).

Theorem 4.68 (Chain rule). *Assume that (O, E), (O', E') and (O'', E'') are local sc-models and that $f : O \to O'$ and $g : O' \to O''$ are sc^1. Then $g \circ f : O \to O''$ is sc^1 and*

$$T(g \circ f) = (Tg) \circ (Tf).$$

Moreover, if f, g are sc^k, then so is $g \circ f$. The same applies for sc^∞.

The following remark explains how the current account is related to [25–27].

Remark 4.69. In the series of papers [25–27], we developed a generalized Fredholm theory in a slightly more restricted situation, which, however, is more than enough for the applications. Namely, rather than considering sc-smooth retractions and sc-smooth retracts, splicings and open subsets of splicing cores were considered, which one can view as a special case. A splicing consists of an open subset V in some sc-Banach space W and a family of bounded linear projections $\pi_v : E \to E, v \in V$, where E is another sc-Banach space, such that the map

$$V \oplus E \to E : (v, e) \to \pi_v(e)$$

is sc-smooth. Then the associated splicing core is K, defined by

$$K = \{(v, e) \in V \oplus E \,|\, \pi_v(e) = e\}.$$

Clearly, $V \oplus E$ is an open subset in $W \oplus E$ and $r(v, e) := (v, \pi_v(e))$ is an sc-smooth retraction. The associated retract is, of course, the splicing core K. If O is an open subset of K, then we know that it is again an sc-smooth retract. Let us note that in all our applications, the retractions are obtained from splicings. The above modifications have been implemented in [31, 32].

We demonstrate next how the definition of a manifold can be generalized. Let Z be a metrizable topological space. A chart for Z is a tuple $(\varphi, U, (O, E))$, where $\varphi : U \to O$ is a homeomorphism and (O, E) is a local sc-model. We say that two such charts are sc-smoothly compatible provided that

$$\psi \circ \varphi^{-1} : \varphi(U \cap V) \to \psi(U \cap V)$$

is sc-smooth, and similarly for $\varphi \circ \psi^{-1}$. Here $(\psi, V, (P, F))$ is the second chart. Note that the sets $\varphi(U \cap V)$ and $\psi(U \cap V)$ are sc-smooth retracts for sc-smooth retractions defined on open sets in E and F, respectively. An sc-smooth atlas for Z consists of a family of sc-smoothly compatible charts whose domains cover Z. Two sc-smooth atlases are compatible provided their union is an sc-smooth atlas. This defines an equivalence relation.

Definition 4.70. *Let Z be a metrizable space. An sc-smooth structure on Z is given by an sc-smooth atlas. Two sc-smooth structures are equivalent if the union of the two associated atlases defines again an sc-smooth structure. An M-polyfold is a metrizable space Z together with an equivalence class of sc-smooth structures.*

We note that these spaces have a natural filtration $Z_0 \supset Z_1 \supset Z_2 \supset \cdots$. The points in Z_i should be viewed as the points of some regularity i. The sc-smooth spaces are a very general type of space on which one can define sc-smooth functions.

It is possible to generalize many of the constructions from differential geometry to these spaces. If we have an sc-smooth partition of unity, we can define Riemannian metrics and consequently a curvature tensor. Note, however, that curvature would only be defined at points of regularity at least 2. The existence of an sc-smooth partition of unity depends on the sc-structure.

The tangent space at a point of level at least 1 is defined in the same way as one defines tangent spaces for Banach manifolds; see [37]. Namely, one considers tuples $(z, \varphi, U, (O, E), h)$, where $z \in Z_1$ and $(\varphi, U, (O, E))$ is a chart such that $z \in U$ and $h \in T_{\varphi(z)}O$. Two such tuples, with say the second being $(z', \varphi', U', (O', E'), h')$, are said to be equivalent provided that $z = z'$ and $T(\varphi' \circ \varphi^{-1})(\varphi(z))h = h'$. An equivalence class $[(z, \varphi, U, (O, E), h)]$ is then, by definition, a tangent vector at z. The tangent space at $z \in Z_1$ is denoted by $T_z Z$ and we define TZ as

$$TZ = \bigcup_{z \in Z_1} \{z\} \times T_z Z.$$

One can show that TZ has a natural M-polyfold structure such that the natural map $TZ \to Z^1$ is sc-smooth; see [31].

Example 4.71. *Consider the metrizable space Z given as the subspace of $\mathbb{R}^2$ defined by*

$$Z = \{(s, t) \in \mathbb{R}^2 \mid t = 0 \text{ if } s \leq 0\}.$$

Then Z admits the structure of an M-polyfold. In order to see this, one constructs a topological embedding into $\mathbb{R} \oplus L^2(\mathbb{R})$, where $L^2(\mathbb{R})$ has the previously introduced sc-structure, in such a way that the image is an sc-smooth retract. Here the idea of an sc-smooth splicing comes in handy! Take a smooth, compactly supported map $\beta : \mathbb{R} \to [0, \infty)$ with $\int \beta(t)^2 \, ds = 1$. Denote by f_s, for $s \in (0, \infty)$, the unit-length element in L^2 defined by

$$f_s(t) = \beta(t + e^{1/s}).$$

For $s \in (-\infty, 0]$, we define $f_s = 0$, and denote by π_s the L^2-orthogonal projection onto the subspace spanned by f_s. Then a somewhat lengthy computation shows that

$$r : \mathbb{R} \oplus L^2 \to \mathbb{R} \oplus L^2 : r(s, u) = (s, \pi_s(u))$$

is an sc-smooth retraction, with obvious image O being

$$\{(s, t \cdot f_s) \mid (s, t) \in \mathbb{R}^2\}.$$

Hence $(O, \mathbb{R} \oplus L^2)$ is a local sc-model. We note that it has varying dimension. The map

$$Z \to \mathbb{R} \oplus L^2 : (s, t) \to (s, t \cdot f_s)$$

is a homeomorphic embedding onto O. The map is clearly continuous and injective and has image O. Define $\mathbb{R} \oplus L^2 \to \mathbb{R}^2$ by

$$(s, x) \to \left(s, \int_{\mathbb{R}} x(t) f_s(t) \, dt \right).$$

This map is continuous, and its restriction to O is the inverse of the previously defined map. Hence we obtain the structure of an M-polyfold on Z. This gives us the first example of a finite-dimensional space, with varying dimension, that has a generalized manifold structure. We also note that the induced filtration is constant, so that a tangent space is defined at all points. This is due to the fact that the local model O lies entirely in the smooth part of $\mathbb{R} \oplus L^2$. It is instructive to study sc-smooth curves $\phi : (-\varepsilon, \varepsilon) \to O$ satisfying $\phi(0) = (0, 0)$. Modifications of the above construction allow us to put smooth structures on the spaces shown in Fig. 4.1. The M-polyfold does not allow an sc-smooth embedding into any $\mathbb{R}^N$, since then it would have be a smooth manifold by [4]. However, as seen in the construction, it can be sc-smoothly embedded into an infinite-dimensional space.

4.3.3 Strong bundles

The notion of a strong bundle is designed to provide additional structures in the Fredholm theory that guarantee a compact perturbation and transversality theory. The crucial point is the fact that there will be a well-defined vector space of perturbations that have certain compactness properties. On the other hand, these perturbations are plentiful enough to allow for different versions of Sard–Smale type theorems [56] in the Fredholm theory.

Let us start with a non-symmetric product $U \lhd F$, where U is an open subset in some sc-Banach space E, and F is also an sc-Banach space. By definition, as a set $U \lhd F$ is the product $U \times F$, but in addition it has a double filtration

$$(U \lhd F)_{m,k} = U_m \oplus F_k$$

defined for all pairs (m, k) satisfying $0 \leq k \leq m + 1$. We view $U \lhd F \to U$ as a bundle with base space U and fiber F, where the double filtration has the interpretation that above a point $x \in U$ of regularity m it makes sense to talk about fiber regularity of a point (x, h) up to order k provided $k \leq m + 1$. At this point, it is not clear why one introduces this non-symmetric product coming with a non-symmetric double filtration. We refer the reader to the later Example 4.74, explaining why it is introduced.

Given $U \lhd F$, we might consider the associated sc-spaces $U \oplus F$ and $U \oplus F^1$. Of interest for us are the maps

$$\Phi : U \lhd F \to V \lhd G$$

of the form

$$\Phi(u, h) = (\varphi(u), \phi(u, h)),$$

which are linear in h.

Definition 4.72. *We say that the map Φ as described above is of class $sc^0_\lhd$ provided that it induces sc^0-maps $U \oplus F^i \to V \oplus G^i$ for $i = 0, 1$.*

We define the tangent $T(U \lhd F)$ by

$$T(U \lhd F) = (TU) \lhd (TF).$$

Note that the order of the factors is different from the order in $T(U \oplus F)$. One has to keep this in mind. Indeed,

$$T(U \lhd F) = U_1 \oplus E \oplus F_1 \oplus F \text{ and } T(U \oplus F) = U_1 \oplus F_1 \oplus E \oplus F.$$

Definition 4.73. *A map $\Phi : U \lhd F \to V \lhd G$ is of class $sc^1_\lhd$ provided that the maps $\Phi : U \oplus F^i \to V \oplus G^i$ for $i = 0, 1$ are sc^1. Taking the tangents of the latter, gives, after rearrangement, the $sc^0_\lhd$-map*

$$T\Phi : (TU) \lhd (TF) \to (TV) \lhd (TG).$$

Iteratively, we can define what it means that a map is $sc^k_\lhd$ for $k = 1, 2, \ldots$ and we can also define $sc_\lhd$-smooth maps.

Given $U \lhd F \to U$, an sc-smooth section f is a map of the form $x \to (x, \bar{f}(x))$ such that the induced map $U \to U \oplus F$ is sc-smooth. In particular, f is "horizontal" with respect to the filtration, i.e., a point on level m is mapped to a point of bi-level (m, m). This can be considered as a convention, and it is precisely this convention that is responsible for the filtration constraint $k \leq m + 1$. There is another class of sections called sc^+-sections. These are sc-smooth sections of $U \lhd F \to U$ that induce sc-smooth maps $U \to U \oplus F^1$. In particular, if s is an sc^+-section of $U \lhd F \to U$ and $s(x) = (x, \bar{s}(x))$ for $x \in U_m$, then $\bar{s}(x) \in F_{m+1}$. This type of section will be important for the perturbation theory. Indeed, it is a kind of compact perturbation theory, since the inclusion $F_{m+1} \to F_m$ is compact. We give an example before we generalize an earlier discussion about retracts and retractions to bundles of the type $U \lhd F \to U$.

Example 4.74. *Let us denote by E the Sobolev space $H^1(S^1, \mathbb{R}^n)$ of loops. We define an sc-structure by $E_m = H^{1+m}(S^1, \mathbb{R}^n)$. Further, we define $F = L^2(S^1, \mathbb{R}^n) = H^0(S^1, \mathbb{R}^n)$, which we filter via $F_m = H^m(S^1, \mathbb{R}^n)$. Finally, we introduce $E \triangleleft F \to E$. Then we can view the map $f : x \to \dot{x}$ as an sc-smooth section of $E \triangleleft F \to E$. In particular, f maps E_m into $E_m \oplus F_m$. We observe that the filtration of F is picked in such a way that the first-order differential operator $x \to \dot{x}$ is an sc-smooth section; in particular, it is horizontal, i.e., the choices are made in such a way that they comply with our convention that sc-smooth sections are index-preserving. The map $x \to x$ can be viewed as an sc^+-section. Then $x \to \dot{x} + x$ is an sc-smooth section obtained from the sc-smooth section $x \to \dot{x}$ via the perturbation by an sc^+-section. Consider now a smooth vector bundle map*

$$\Phi : \mathbb{R}^n \oplus \mathbb{R}^n \to \mathbb{R}^n \oplus \mathbb{R}^n$$

of the form

$$\Phi(x, h) = (\varphi(x), \phi(x)h),$$

where $\varphi : \mathbb{R}^n \to \mathbb{R}^n$ is a diffeomorphism and for every $x \in \mathbb{R}^n$ the map $\phi(x) : \mathbb{R}^n \to \mathbb{R}^n$ is a linear isomorphism. Then we define for $(x, h) \in E \oplus F$ the element $\Phi_*(x, h)(t) = (\varphi(x(t)), \phi(x(t))h(t))$. Note that if $x \in E_m$ and $h \in F_k$ for $k \leq m + 1$, then $\Phi_*(x, h) =: (y, \ell)$ satisfies $y \in E_m$ and $\ell \in F_k$. However, if $x \in E_m$ and $y \in F_k$ for some $k > m + 1$, we cannot conclude that $\ell \in F_k$. We can only say that $\ell \in F_{m+1}$. Now, one easily verifies that

$$\Phi_* : E \triangleleft F \to E \triangleleft F$$

is $sc_\triangleleft$-smooth. This justifies our constraint $k \leq m + 1$ for the double filtration.

Definition 4.75. *An $sc_\triangleleft^\infty$-retraction is an $sc_\triangleleft$-smooth map*

$$R : U \triangleleft F \to U \triangleleft F$$

with the property $R \circ R = R$.

Of course, R has the form $R(u, h) = (r(u), \phi(u, h))$, with r being an sc-smooth retraction and $\phi(u, h)$ linear in the fiber. Given R, we can define its image $K = R(U \triangleleft F)$ and $O = r(U)$. Then we have a natural projection map

$$p : K \to O.$$

We may view this as the local model for a strong bundle. Observe that K has a double filtration and p maps points of regularity (m, k) to points of regularity m.

Definition 4.76. *The tuple $(K, E \triangleleft F)$, where K is a subset of $E \triangleleft F$, such that there exists an $sc_\triangleleft^\infty$-retraction R defined on $U \triangleleft F$, where $U \subset E$ is open and $K = R(U \triangleleft F)$, is called a local strong bundle model.*

Starting with $(K, E \triangleleft F)$, we have the projection $K \to E$ and denote its image by O and the induced map by $p : K \to O$. We can define $T(K, E \triangleleft F)$ by

$$T(K, E \triangleleft F) = (TK, TE \triangleleft TF),$$

where TK is the image of TR. As before, we can show that the definition does not depend on the choice of R.

Now we are in the position to define the notion of a strong bundle. Let $p : W \to X$ be a surjective continuous map between two metrizable spaces such that for every $x \in X$, the space $W_x := p^{-1}(x)$ comes with the structure of a Banach space. A strong bundle chart is a tuple $(\Phi, p^{-1}(U), E \triangleleft F))$, where $\Phi : p^{-1}(U) \to K$ is a homeomorphism, covering a homeomorphism $\varphi : U \to O$, which between each fiber is a bounded linear operator

$$
\begin{array}{ccc}
p^{-1}(U) & \xrightarrow{\ \Phi\ } & K \\
\ \downarrow{\scriptstyle p} & & \downarrow \\
U & \xrightarrow{\ \varphi\ } & O
\end{array}
$$

We call two such charts $sc_\triangleleft$-smoothly equivalent if the associated transition maps are $sc_\triangleleft$-smooth. We can define the notion of a strong bundle atlas and also the notion of equivalence of two such atlases.

Definition 4.77. *Let $p : W \to X$ be as described before. A strong bundle structure for p is given by a strong bundle atlas. Two strong bundle structures are equivalent if the associated atlases are equivalent. Finally, p equipped with an equivalence class of strong bundle atlases is called a strong bundle.*

Let us observe that a strong bundle $p : W \to X$ admits a double filtration $W_{m,k}$ with $0 \leq k \leq m + 1$. By forgetting part of this double filtration, we observe that $W(0)$, which is W filtered by $W(0)_m := W_{m,m}$, has in a natural way the structure of an M-polyfold. The same is true for $W(1)$, which is the space $W_{0,1}$ equipped with the filtration $W(1)_m := W_{m,m+1}$. Obviously, the maps $p : W(i) \to X$ for $i = 0, 1$ are sc-smooth.

The previously introduced notions of sc-smooth sections and sc^+-sections for $U \triangleleft F \to U$ generalize as follows.

Definition 4.78. *Let $p : W \to X$ be a strong bundle over the M-polyfold X (without boundary).*

(i) An sc-smooth section of the strong bundle p is an sc^0-map $s : X \to W$ with $p \circ s = Id_X$ such that $s : X \to W(0)$ is sc-smooth. The vector space of all such sections is written as $\Gamma(p)$.

(ii) An sc^+-section section of the strong bundle p is an sc^0-map $s : X \to W(1)$ with $p \circ s = Id_X$, which in addition is sc-smooth. The vector space of sc^+-sections is denoted by $\Gamma^+(p)$.

In some sense, sc^+-sections are compact perturbations, since the inclusion map $W(1) \to W(0)$ is fiber-wise compact. They are very important for the perturbation theory.

4.3.4 A special class of sc-smooth germs

The next goal is to define a suitable notion of Fredholm section of a strong bundle. The basic fact about the usual Fréchet differentiability is the following. If $f : U \to F$ is a smooth map (in the usual sense) between an open neighborhood U of $0 \in E$ with target the Banach space F, and satisfying $f(0) = 0$, then we can describe the solution set of $f = 0$ near 0 by an implicit function theorem provided that $df(0)$ is surjective and the kernel of $df(0)$ splits, i.e., has a topological linear complement. So smoothness and some properties of the linearized operator at a solution always give us qualitative knowledge about the solution set near 0. On the other hand, $f : U \to F$ being only sc-smooth, $df(0)$ being surjective and its kernel having an sc-complement are not enough to conclude much about the solution space near 0. However, as we shall see, there is a large class of sc-smooth maps for which a form of the implicit function theorem holds. In applications, the class is large enough to explain gluing constructions (à la Taubes and Floer) as smooth implicit function theorems in the sc-world.

One of the issues that has to be addressed at some point is the fact that the spaces we are concerned with have locally varying dimensions. Although this might seem like a major issue, it will turn out that there is a simple way to deal with such problems. It is a crucial observation that in applications, base and fiber dimension change coherently. The sc-formalism incorporates this with a minimum amount of technicalities. We should remark that our presentation is slightly more general than that given in [26].

Let us begin with some notation. As usual, E is an sc-Banach space. We shall write $\mathcal{O}(E, 0)$ for an unspecified nested sequence $U_0 \supset U_1 \supset U_2 \supset \cdots$, where every U_i is an open neighborhood of $0 \in E_i$. Note that this differs from previous notation where $U_i = E_i \cap U$. When we are dealing with germs, we always have the new definition in mind. An sc-smooth germ

$$f : \mathcal{O}(E, 0) \to F$$

is a map defined on U_0 such that for points $x \in U_1$, the tangent map $Tf : U_1 \oplus E_0 \to TF$ is defined, which again is a germ

$$Tf : \mathcal{O}(TE, 0) \to TF.$$

We introduce a basic class $\mathfrak{C}_{\text{basic}}$ of germs of maps as follows.

Definition 4.79. *An element in $\mathfrak{C}_{\text{basic}}$ is an sc-smooth germ*

$$f : \mathcal{O}(\mathbb{R}^n \oplus W, 0) \to (\mathbb{R}^N \oplus W, 0)$$

*for suitable n and N, and an sc-Banach space W, such that the following holds. If $P : \mathbb{R}^N \oplus$
$W \to W$ is the projection, then $P \circ f$ has the form*

$$P \circ f(r, w) = w - B(r, w)$$

for $(r, w) \in U_0 \subset \mathbb{R}^n \oplus W$. Moreover, for every $\varepsilon > 0$ and $m \in \mathbb{N}$, we have

$$\left\| B(r, w) - B(r, w') \right\|_m \le \varepsilon \left\| w - w' \right\|_m$$

for all $(r, w), (r, w') \in U_m$ close enough to $(0, 0)$ on level m.

In [26], the class of basic germs was slightly more general in the sense that it was not required that $f(0) = 0$ in its definition. However, all important results were then proved under the additional assumption that $f(0) = 0$. In the applications to SFT and the other mentioned theories, one can bring the nonlinear elliptic differential operators that occur, even at bubbling-off points (modulo a filling, which is a crucial concept in the poly-fold theory and will be explained shortly), via sc-smooth coordinate changes into the above form; see [30] for Gromov–Witten theory and [12] for the operators in SFT. It is important to note that if f is sc-smooth such that $f(0) = 0$ and $Df(0)$ is sc-Fredholm, it is generally not true that after a change of coordinates f can be pushed forward to an element that belongs to $\mathfrak{C}_{\text{basic}}$.

As shown in [26, 31], basic germs admit something like infinitesimal smooth implicit function theorems near 0 (this is something intrinsic to sc-structures) that for certain maps can be "bound together" to give a local implicit function theorem.

To explain this, assume that $U \subset E$ is an open neighborhood of 0 and $f : U \to F$ is sc-smooth having the following properties, where $U_i = E_i \cap U$:

(i) $f(0) = 0$ and $Df(0)$ is a surjective sc-Fredholm operator.

(ii) f is regularizing. This means that if $x \in U_m$ and $f(x) \in F_{m+1}$, then $x \in U_{m+1}$.

(iii) Viewing f as a section of $U \lhd F \to U$, near every smooth point x, and for a suitable sc^+-section with $s(x) = f(x)$, the germ

$$f - s : \mathcal{O}(E, 0) \to F$$

is conjugated to a basic germ.

Under these conditions, there is a local implicit function theorem near 0, which guarantees a local solution set of dimension given by the Fredholm index of $Df(0)$ at 0, and in addition guarantees a natural manifold structure on this solution set.

The infinitesimal implicit function theorem refers to the following phenomenon for basic germs. If $f \in \mathfrak{C}_{\text{basic}}$, then $Pf(a, w) = w - B(a, w)$, where B is a family of contractions on every level m near $(0, 0)$. Hence, using Banach's fixed-point theorem, we find a germ δ_m solving $\delta_m(a) = B(a, \delta_m(a))$ on level m for a near 0. By uniqueness, a solution on level m also solves the problem on lower levels. This implies that we have a solution germ

$a \to (a, \delta(a))$ of $Pf(a,w) = 0$. The infinitesimal sc-smooth implicit function theorem gives the non-trivial fact that the germ

$$\delta : \mathcal{O}(\mathbb{R}^n, 0) \to (W, 0)$$

is an sc-smooth germ; see [26, 31].

In summary, as we shall discuss in more detail later, if we have a regularizing sc-smooth section that around every smooth point is conjugated mod a suitable sc$^+$-section to a basic germ, then the "infinitesimal" implicit function theorems around points y near x combine together to give a "local" implicit function theorem near a point x where the linearization is surjective. We refer the reader to [31] for a comprehensive discussion.

4.3.5 sc-Fredholm sections

Assume next that $p : K \to O$ is a strong local bundle, i.e., $(K, E \triangleleft F)$ is a local strong bundle model. Suppose $f : \mathcal{O}(O, x) \to K$ is a germ, which we shall write as $[f, x]$.

Definition 4.80. *A filling for the germ $[f, x]$ consists of the following data:*

1. *An sc-smooth germ $\bar{f} : \mathcal{O}(E, x) \to F$.*
2. *A choice of strong bundle retraction $R : U \triangleleft F \to U \triangleleft F$ such that K is the image of R.*

Viewing f as a map $O \to F$ such that $\phi(y)f(y) = f(y)$, where $R(y, h) = (r(y), \phi(y)h)$, we assume that the data satisfies the following properties:

1. *$\bar{f}(y) = f(y)$ for all $y \in O$ near x.*
2. *$\bar{f}(y) = \phi(r(y))\bar{f}(y)$ for y near x in U implies that $y \in O$.*
3. *The linearization of the map*

$$y \to (Id - \phi(r(y)))\bar{f}(y)$$

at x restricted to $\ker(Dr(x))$ defines a linear topological isomorphism $\ker(Dr(x)) \to \ker(\phi(x))$.

The germ $[f, x]$ is said to be fillable provided there exists a germ of a strong bundle map Φ, covering a germ of a (local) sc-diffeomorphism φ, such that the push-forward germ $[\Phi_*(f), \varphi(x)]$ has a filling. A filled version of $[f, x]$ is an sc-smooth germ $[\bar{g}, \bar{x}]$ obtained as a filling of a suitable push-forward.

In the definition, the meaning that Φ is a germ of a strong bundle map covering φ is the following. For the given $(K, U \triangleleft F)$, $p : K \to O$, with $x \in O \subset U$, there exists $(K', U' \triangleleft F')$, $p' : K' \to O'$, with $x' \in O' \subset E'$ and open neighborhoods $x \in V \subset O$, $x' \in V' \subset O'$, such that the following is a commutative diagram associated to a strong bundle isomorphism:

$$p^{-1}(V) \xrightarrow{\quad \Phi \quad} (p')^{-1}(V')$$

$$p \downarrow \qquad\qquad p' \downarrow$$

$$V \xrightarrow{\quad \varphi \quad} V'$$

Moreover, the sizes of V and then of V' are unspecified small, but fixed. Also, in this case, $V_i = O_i \cap V$. For the solution germs, the neighborhoods in higher regularity shrink.

If $[f, x]$ has a filling $[\bar{f}, x]$, then the local study of $f(y) = 0$ with $y \in O$ near x is equivalent to the local study of $\bar{f}(y) = 0$, where $y \in U$ close to x. Let us note that if $f(x) = 0$, then the linearization $f'(x) : T_x O \to K_x$ has the same kernel as $\bar{f}'(x) : T_x U \to F_x$ and the cokernels are naturally isomorphic.

Definition 4.81. *If f is an sc-smooth section of a strong M-polyfold bundle $p : W \to X$ (with $\partial X = \varnothing$), and x is a smooth point, we say that the germ $[f, x]$ admits a filled version, provided that, in a suitable local coordinate representation, $[f, x]$ admits a filled version as defined in Definition 4.80. We always may assume that the filled version has the form $g : \mathcal{O}(E, 0) \to F$.*

We recall the notion of a regularizing section, which we have already mentioned.

Definition 4.82. *Let $p : W \to X$ be a strong bundle over the M-polyfold X (without boundary) and f an sc-smooth section. We say that f is regularizing provided that for a point $x \in X$, the assertion $f(x) \in W_{m,m+1}$ implies that $x \in X_{m+1}$.*

Note that for a regularizing section f, a solution x of $f(x) = 0$ belongs necessarily to X_∞. If f is regularizing and $s \in \Gamma^+(p)$, then $f + s$ is regularizing. Now we come to the crucial definition.

Definition 4.83. *We call the sc-smooth section f of the strong bundle $p : W \to X$ over an M-polyfold (with $\partial X = \varnothing$) an sc-Fredholm section provided that f is regularizing and, around every smooth point x, the germ $[f, x]$ has a filled version $[g, 0]$ such that for a suitable germ of sc^+-section s with $s(0) = g(0)$, the germ $[g - s, 0]$ is conjugated to an element in $\mathfrak{C}_{\text{basic}}$. We denote the collection of all sc-Fredholm sections of p by $\mathcal{F}(p)$.*

Remark 4.84. An sc-Fredholm section according to the above definition is slightly more general than the sc-Fredholm sections defined in [26]. An additional advantage of the current definition is the stability result saying that, given an sc-Fredholm section for $p : W \to X$ and an sc^+-section s, then $f + s$ is an sc-Fredholm section for $p : W \to X$. With the version given in [26], one can only conclude that $f + s$ is an sc-Fredholm section of $p^1 : W^1 \to X^1$. In applications, the difference is only "academic." However, as far as a presentation is concerned, this new version is more pleasant—see [31].

The following stability result is crucial for the perturbation theory and rather tautological. In the setup of [26], it was a non-trivial theorem. However, some of the burden is now moved to the implicit function theorem—see [31] for the proofs in this new setup.

Theorem 4.85 (Stability). *Let* $p : W \to X$ *a strong bundle over the M-polyfold X. Then, given* $f \in \mathcal{F}(p)$ *and* $s \in \Gamma^+(p)$, *we have that* $f + s \in \mathcal{F}(p)$.

Fredholm sections allow for an implicit function theorem.

Theorem 4.86. *Assume that* $p : W \to X$ *is a strong bundle over the M-polyfold X (without boundary). Let* f *be a sc-Fredholm section and* x *a smooth point such that* $f(x) = 0$ *and* $f'(x) : T_x X \to W_x$ *is surjective. Then the solution set near* x *carries in a natural way the structure of a smooth manifold with dimension given by the Fredholm index of* $f'(x)$. *In addition, there exists an open neighborhood* V *of* x *such that for every* $y \in V$ *with* $f(y) = 0$, *the linearization* $f'(y)$ *is surjective. Moreover, its kernel can be identified with the tangent spaces of the solution set at* y.

Finally, we introduce the notion of an auxiliary norm and give a useful compactness result.

Definition 4.87. *Assume that* $p : W \to X$ *is a strong bundle over the M-polyfold X (without boundary). An auxiliary norm for* p *is a map* $N : W \to \mathbb{R}^+ \cup \{+\infty\}$ *having the following properties:*

   (i) The restriction $N|W_{0,1}$ *is real-valued and continuous, and on* $W \setminus W_{0,1}$ *the map* N *takes the value* ∞.

   (ii) N restricted to any fiber of $p : W_{0,1} \to X$ *is a complete norm.*

   (iii) If $(h_k) \subset W_{0,1}$ *is a sequence such that* $p(h_k) \to x_0$ *in X, and* $N(h_k) \to 0$, *then* $h_k \to 0_{x_0}$ *in* $W_{0,1}$.

The existence of an auxiliary norm can be establishes using continuous partitions of unity.

Proposition 4.88. *Given a strong bundle* $p : W \to X$ *over the M-polyfold X (without boundary), there exists an auxiliary norm N. For two given auxiliary norms* N_1 *and* N_2, *there exist continuous maps* $f_1, f_2 : X \to (0, \infty)$ *such that*

$$f_1 \cdot N_1 \leq N_2 \leq f_2 \cdot N_1.$$

Now we are in a position to state a useful compactness result for sc-Fredholm sections that is important for the perturbation theory; see [31].

Theorem 4.89. *Let* $p : W \to X$ *be a strong bundle over the M-polyfold X (without boundary). Suppose that* f *is an sc-Fredholm section for which* $f^{-1}(0)$ *is compact.*

   (i) Given an auxiliary norm N, there exists an open neighborhood U of $f^{-1}(0)$ *such that for every* sc^+-section $s \in \Gamma^+(p)$ *with support in U, satisfying* $N(s(x)) \leq 1$ *for all* $x \in X$, *the set* $(f + s)^{-1}(0)$ *is compact.*

   (ii) If X admits sc-smooth partitions of unity, then we find for every $\varepsilon \in (0, 1]$ *an* sc^+-*section* s *with support in U such that* $N(s(x)) < \varepsilon$ *for all* x, *and the set* $M = (f + s)^{-1}(0)$ *has the structure of a compact smooth manifold without boundary,*

such that the linearization $f'(m) : T_m M \to W_m$ for all $m \in M$ is surjective and the tangent space $T_m M$ can be canonically identified with $\ker(f'(m))$.

4.4 Polyfold structures and consequences

At this point, we have discussed a smooth theory for the category of stable Riemann surfaces based on the construction of good uniformizers, and a topological theory associated to the category of stable maps. In order to transform the latter discussion into one taking place in a smooth world, we have generalized the finite- or infinite-dimensional classical differential geometry to a more general sc-smooth differential geometry and described some of the aspects of an associated nonlinear functional analysis. The differential geometric/nonlinear functional analytic theory is discussed in great detail in [31, 32] and gives many more ideas about the framework outlined in Section 4.3. In the following, we show how to use this theory to equip certain categories with sc-smooth structures and illustrate the ideas with the stable map example.

4.4.1 Polyfold structures on certain categories

We start with a useful definition.

Definition 4.90. *A good category with metrizable orbit space is given by a pair (C, T), where C is a category with the following properties:*

(i) Between any two objects, there are only finitely many morphisms and every morphism is an isomorphism.

(ii) The orbit space $|C|$ is a set.

Moreover, T is a metrizable topology on $|C|$. We call (C, T) a GCT (G = good, C = category, T = topology).

From our discussions in the previous sections, the categories $\mathcal{R}$ and $\mathcal{S}^{3,\delta_0}(Q, \omega)$ are GCTs. Precisely for such categories, a construction of type $(F, \mathbf{M})$ is important and very often exists naturally if C is a category coming from geometric considerations. Next we give a precise definition of the type of previously discussed constructions in the polyfold framework.

Definition 4.91. *Let C be a GCT. A good uniformizer for C around an object c with automorphism group G, written as*

$$\Psi : G \ltimes O \to C,$$

consists of an M-polyfold O with an sc-smooth action of G on O, where $G \ltimes O$ is the associated translation groupoid and Ψ is a functor with the following properties:

(i) Ψ is full and faithful and injective on objects.

(ii) There exists a $q_0 \in O$ with $\Psi(q_0) = c$.

(iii) *Passing to orbit spaces,* $|\Psi| : {}_G\backslash O \to |\mathcal{C}|$ *is a homeomorphism onto an open neighborhood of* $|\alpha|$.

As in the continuous case, we can define for two good uniformizers the transition set $\mathbf{M}(\Psi, \Psi')$, and we have the usual structural maps. Here comes the crucial definition.

Definition 4.92. *Let $\mathcal{C}$ be a GCT. A polyfold structure for $\mathcal{C}$ is a pair $(F, \mathbf{M})$, where $F : \mathcal{C} \to \text{SET}$ is a functor associating to an object c a set $F(c)$ of good uniformizers and $\mathbf{M}$ associates to two good uniformizers, say $\Psi \in F(c)$ and $\Psi' \in F(c')$, an M-polyfold structure to the transition set $\mathbf{M}(\Psi, \Psi')$, such that all structural maps are sc-smooth and the source and target map are local sc-diffeomorphisms.*

Instead of denoting the polyfold structure by $(F, \mathbf{M})$, we just write F, i.e., $F \equiv (F, \mathbf{M})$. Let us emphasize that $F(c)$ is a set of good uniformizers, i.e., a specifically picked collection of good uniformizers constructed by a given procedure. Since classically smooth manifolds are in particular M-polyfolds, we see that the constructions associated to $\mathcal{R}$ equip it with a polyfold structure.

Remark 4.93. There are several useful points one should make:

(a) Note that the category $\mathcal{C}$ can be recovered from F. First of all, since F is defined on $\mathcal{C}$, we know all the objects of $\mathcal{C}$. For two objects c and c', we pick $\Psi \in F(c)$ and $\Psi' \in F(c')$ and let $q \in O$, $q' \in O'$, with $F(q) = c$ and $F(q') = c'$. Then we can identify the elements (q, ϕ, q') in $\mathbf{M}(\Psi(q), \Psi'(q'))$ with the morphisms in the set $\text{mor}_{\mathcal{C}}(c, c')$ via $(q, \phi, q') \to \phi$.

(b) We also note that having a polyfold structure on $\mathcal{C}$, there exists a functor

$$\text{reg} : \mathcal{C} \to \mathbb{N}_0 \cup \{\infty\} =: \mathbb{N}_0^\infty$$

that associates to an object its regularity. Here $\mathbb{N}_0^\infty$ has as morphisms only the identities. Given an object c, we pick $\Psi \in F(c)$ and $q_0 \in O$ such that $\Psi(q_0) = c$. Then we define

$$\text{reg}(c) = \sup\{k \in \mathbb{N}_0 \mid q_0 \in O_k\}.$$

This is well defined, independent of the choices involved and a morphism invariant. We can define the full subcategories $\mathcal{C}_r$ of $\mathcal{C}$ associated to the objects c with $\text{reg}(c) \geq r$.

(c) Given a polyfold structure F for $\mathcal{C}$, we consider $\mathcal{C}_1$ and define $F^1 : \mathcal{C}_1 \to \text{SET}$ by

$$F^1(c) = \{\Psi^1 : G \ltimes O^1 \to \mathcal{C}_1 \mid \Psi \in F(c)\}.$$

One can equip $|\mathcal{C}_1|$ with a uniquely determined metrizable topology that makes every element in $F^1(c)$ a good uniformizer. Then one can show that F^1 defines a polyfold structure on $\mathcal{C}_1$. We shall denote by $\mathcal{C}^1$ the category $\mathcal{C}_1$ equipped with the polyfold structure.

We also need a bundle version. As in Section 4.2.3, we denote by BAN_G the category whose objects are Banach spaces and whose morphisms are invertible topological linear isomorphisms. Suppose we are given a category $\mathcal{C}$ and a functor

$$\mu : \mathcal{C} \to \mathrm{BAN}_G.$$

Then we can build a new category $\mathcal{E} = \mathcal{E}_\mu$ whose objects are pairs (c, h), where c is an object in $\mathcal{C}$ and h is a vector in $\mu(c)$. The morphisms $(c, h) \to (c', h')$ are the lifts of the morphisms $\phi : c \to c'$, namely,

$$(\phi, \mu(\phi), h) : (c, h) \to (c', h')$$

provided that $s(\phi) = c, t(\phi) = c'$ and $\mu(\phi)(h) = h'$. We shall abbreviate $\widehat{\phi} = (\phi, \mu(\phi))$, which is a linear topological isomorphism

$$\widehat{\phi} : P^{-1}(s(\phi)) \to P^{-1}(t(\phi)).$$

We shall denote the class of all morphisms for the category $\mathcal{E}_\mu$ by $\boldsymbol{\mathcal{E}}_\mu$. Since $|\mathcal{C}|$ is a set, the same is true for $|\mathcal{E}|$. We denote by $P = P_\mu : \mathcal{E}_\mu \to \mathcal{C}$ the functor that on objects $(c, h) \to c$ and on morphisms $(\phi, \mu(\phi), h) \to \phi$. We also have the source map $s : \boldsymbol{\mathcal{E}}_\mu \to \mathcal{E}_\mu$ defined by $s(\phi, \mu(\phi), h) = (s(\phi), h)$ and the target map $t(\phi, \mu(\phi), h) = (t(\phi), \mu(\phi)(h))$.

Definition 4.94. *A bundle GCT is given by a tuple $(\mathcal{C}, \mu, \mathcal{T}_\mu)$, where $\mathcal{C}$ is a GCT, $\mu : \mathcal{C} \to \mathrm{BAN}_G$ is a functor and $\mathcal{T}_\mu$ is a metrizable topology on $|\boldsymbol{\mathcal{E}}_\mu|$, such that $|P_\mu| : |\boldsymbol{\mathcal{E}}_\mu| \to |\mathcal{C}|$ is continuous and open.*

We shall introduce the notion of strong bundle uniformizers. The strong bundle uniformizers are build on strong bundles $p : K \to O$ over M-polyfolds, equipped with an action of a finite group G acting by sc-smooth strong bundle isomorphisms inducing an action of G on O by sc-diffeomorphisms, such that p is equivariant.

Definition 4.95. *A good strong bundle uniformizer for the bundle GCT $(\mathcal{C}, \mu, \mathcal{T}_\mu)$ around an object c in $\mathcal{C}$ is a functor $\bar{\Psi} : G \ltimes K \to \mathcal{E}$ covering a functor $\Psi : G \ltimes O \to \mathcal{C}$ such that the following hold:*

(i) *The diagram*

$$
\begin{array}{ccc}
G \ltimes K & \xrightarrow{\ \bar{\Psi}\ } & \mathcal{E}_\mu \\
{\scriptstyle p}\downarrow & & \downarrow{\scriptstyle P_\mu} \\
G \ltimes O & \xrightarrow{\ \Psi\ } & \mathcal{C}
\end{array}
$$

is commutative and $\Psi(q_0) = c$ for some $q_0 \in O$.

(ii) *$\bar{\Psi}$ is full and faithful, and injective on objects.*

(iii) $|\bar{\Psi}| : |K| \to |\mathcal{E}_\mu|$ is a homeomorphism onto an open subset of $|\mathcal{E}_\mu|$ of the form $|P|^{-1}(U)$, where $U = |\Psi(O)|$.

(iv) The map Ψ is fiber-wise a topological linear Banach space isomorphism.

We observe that automatically $\Psi : G \ltimes O \to C$ has to be a good uniformizer. Given two strong bundle uniformizers $\bar{\Psi}$ and $\bar{\Psi}'$, we can, similarly to before, define the transition set

$$\mathbf{M}(\bar{\Psi}, \bar{\Psi}') = \{(h, \bar{\Phi}, h') \mid h \in K, \ h' \in K', \ \bar{\Phi} \in \text{mor}(\bar{\Psi}(h), \bar{\Psi}'(h'))\}.$$

Observe that we have a natural map

$$\mathbf{M}(\bar{\Psi}, \bar{\Psi}') \to \mathbf{M}(\Psi, \Psi') : (h, \bar{\Phi}, h') \to (p(h), \Phi, p'(h')),$$

where a fiber has a a natural Banach space structure. We note that s and t in the following diagram are fiber-wise linear (the top row):

$$
\begin{array}{ccccc}
K & \xleftarrow{\ s\ } & \mathbf{M}(\bar{\Psi}, \bar{\Psi}') & \xrightarrow{\ t\ } & K' \\
{\scriptstyle p}\downarrow & & \downarrow & & {\scriptstyle p'}\downarrow \\
O & \xleftarrow{\ s\ } & \mathbf{M}(\Psi, \Psi') & \xrightarrow{\ t\ } & O'
\end{array}
\qquad (4.1)
$$

The main definition is now that of a strong polyfold bundle structure for $(\mathcal{C}, \mu, \mathcal{T}_\mu)$.

Definition 4.96. *A strong polyfold bundle structure for $(\mathcal{C}, \mu, \mathcal{T}_\mu)$ is given by $(\bar{F}, \mathbf{M})$, where $\bar{F} : \mathcal{C} \to$ SET is a functor associating to every object c in $\mathcal{C}$ a set of good strong bundle uniformizers and to every transition*

$$\mathbf{M}(\bar{\Psi}, \bar{\Psi}') \to \mathbf{M}(\Psi, \Psi')$$

a strong bundle structure such that source and target maps define local strong bundle isomorphisms and all structural maps are strong bundle maps—see the diagram 4.1.

4.4.2 Tangent category and differential forms

Assume that $\mathcal{C}$ is a GCT equipped with a polyfold structure $(F, \mathbf{M})$. Then we have a filtration and inclusion functors, since, as a consequence of this construction, we can talk about the regularity of an object

$$\mathcal{C}_\infty \cdots \to \mathcal{C}_{i+1} \to \mathcal{C}_i \cdots \to \mathcal{C}_0 = \mathcal{C}.$$

We can define the tangent category $T\mathcal{C}$ together with a projection functor $T\mathcal{C} \to \mathcal{C}^1$. Here $\mathcal{C}^1$ is the category $\mathcal{C}_1$ with filtration $\mathcal{C}_i^1 := \mathcal{C}_{i+1}$ (and its polyfold structure). Consider tuples $(c, \Psi, (q, h))$, where c is an object in $\mathcal{C}_1$, $\Psi \in F(c)$, say $\Psi : G \ltimes O \to \mathcal{C}$, $\Psi(q) = c$, and $h \in T_q O$. We shall introduce the notion of equivalence of two such tuples. For this, consider a second one, say $(c', \Psi', (q', h'))$. Take suitable open neighborhoods $U(q, 1_c, q') \subset \mathbf{M}(\Psi, \Psi')$, $U(q) \subset O$ and $U(q') \subset O'$ such that that the source and target maps

$$U(q) \xleftarrow{s} U(q, 1_c, q') \xrightarrow{t} U(q')$$

are sc-diffeomorphisms and define

$$L : U(q) \to U(q') : L(p) = (t \circ (s|U(q, 1_c, q'))^{-1})(p). \tag{4.2}$$

We call the two tuples equivalent, written as

$$(c, \Psi, (q, h)) \simeq (c', \Psi', (q', h')),$$

provided that the following holds:

$$c = c' \quad \text{and} \quad TL(q)(h) = (q', h').$$

We denote an equivalence class by $[(c, \Psi, (q, h))]$ and view these as objects in a category denoted by $T\mathcal{C}$. On the object level, we have the projection functor $\tau : T\mathcal{C} \to \mathcal{C}^1$ given by

$$\tau([c, \Psi, (q, h)]) = c.$$

We observe that $\tau^{-1}(c)$ is a Banach space. Given a morphism $\phi : c \to c'$, we can define a topological linear isomorphism

$$T\phi : \tau^{-1}(c) \to \tau^{-1}(c')$$

as follows. Take uniformizers Ψ at c and Ψ' at c' and assume that $\Psi(q) = c$ and $\Psi'(q') = c'$. Similarly as in (4.2), we consider

$$L : U(q) \to U(q') : L(p) = (t \circ (s|U(q, \phi, q'))^{-1})(p)$$

and define

$$T\phi([(c, \Psi, (q, h))]) = [(c', \Psi', TL(q, h)]. \tag{4.3}$$

The morphisms in $T\mathcal{C}$ are given by the tuples

$$\Phi := ([(c, \Psi, (q, h))], T\phi, [(c', \Psi', (q', h'))]),$$

where $\phi : c \to c'$ and $T\phi([(c, \Psi, (q, h))]) = [(c', \Psi', (q', h'))]$. Here

$$s(\Phi) = [(c, \Psi, (q, h))] \text{ and } t(\Phi) = [(c', \Psi', (q', h'))].$$

The projection functor τ on the level of morphisms is defined by

$$\tau(([(c, \Psi, (q, h))], T\phi, [(c', \Psi', (q', h'))]) = \phi.$$

At this point, we have shown the following.

Lemma 4.97. *Let C be a GCT equipped with a polyfold structure $(F, \mathbf{M})$. Then there is a well-defined tangent category TC whose objects are the equivalence classes $[(c, \Psi, (q, h))]$, with c being an object in C^1, $(q, h) \in T_q O$, where $\Psi(q) = c$ and $\Psi \in F(c)$. The morphisms are given by the tuples*

$$\Phi = ([(c, \Psi, (q, h))], T\phi, [(c', \Psi', (q', h'))]),$$

where $T\phi([(c, \Psi, (q, h))]) = [(c', \Psi', (q', h'))]$. In addition, the projection functor $\tau : TC \to C^1$ is defined by $\tau(\Phi) = \phi$ on morphisms and $\tau([(c, \Psi, (q, h))]) = c$ on objects.

We shall show that we can equip TC with a polyfold structure as well. This needs some preparation. Fix an object c_0 in C^1 and pick $\Psi_0 \in F(c_0)$ that is given by

$$\Psi_0 : G \ltimes O \to C.$$

There is an element $q_0 \in O_1$ with $\Psi_0(q_0) = c_0$. For $q \in O_1$, let $c = \Psi_0(q)$ and note that c is an object in C_1. Pick $\Psi \in F(c)$, say $\Psi : G' \ltimes O' \to C$, and $q'_0 \in O'$ with $\Psi(q'_0) = c = \Psi_0(q)$. Then $q'_0 \in O'_1$. Now we us the transition M-polyfold $\mathbf{M}(\Psi_0, \Psi)$ and take open neighborhoods $U(q)$, $U(q'_0)$ and $U(q, 1_c, q'_0)$ such that we have a diagram of sc-diffeomorphisms

$$U(q) \xleftarrow{s} U(q, 1_c, q'_0) \xrightarrow{t} U(q'_0).$$

We consider

$$L : U(q) \to U(q'_0) : p \to L(p) := L = t \circ (s|U(q, 1_c, q'))^{-1}(p)$$

and define for $(q, h) \in T_q O$, which is an object in $G \ltimes TO$,

$$T\Psi_0(q, h) = [(c, \Psi, TL(q, h))],$$

which belongs to $\tau^{-1}(c)$. For a morphism $(g, (q, h)) : (q, h) \to (g * q, g * h)$ in $G \ltimes TO$, we define the morphism

$$T\Psi_0(g, (q, h)) : T\Psi_0(q, h) \to T\Psi_0(g * q, g * h)$$

by

$T\Psi_0(g, (q, h))$

$\quad = ([(c, \Psi, TL(q, h))], T(\Psi_0(g, q)), T(\Psi_0(g, q))([(c, \Psi, TL(q, h))]))$

$\quad = (T\Psi_0(q, h), T(\Psi_0(g, q)), T(\Psi_0(q, q))(T\Psi_0(q, h))).$

Here $T(\Psi_0(g,q))$ is the tangent associated to the morphism $\Psi_0(g,q) : \Psi_0(q) \to \Psi_0(g*q)$ defined as in (4.3).

We define for an object c_0 in $\mathcal{C}^1$.

$$\bar{F}(c_0) = \{T\Psi : G \ltimes TO \to TC \mid \Psi \in F(c_0)\},$$

and obtain a functor $\bar{F} : \mathcal{C}^1 \to \text{SET}$. Then we define

$$TF : TC \to \text{SET} : TF = \bar{F} \circ \tau,$$

i.e., $TF([(c_0, \Psi_0, (q_0, h_0))]) := \bar{F}(c_0)$. Given $T\Psi \in TF(([(c_0, \Psi_0, (q_0, h_0))])$ and $T\Psi' \in ([(c_0', \Psi_0', (q_0', h_0'))])$, we can build $\mathbf{M}(T\Psi, T\Psi')$. The basic result is the following theorem.

Theorem 4.98. *Given a polyfold structure F for the GCT $\mathcal{C}$, there exists an associated natural polyfold structure TF for $TC \to \mathcal{C}^1$ covering the lifted one for $\mathcal{C}^1$ given by F^1; see the diagram (4.4) below.*

More precisely, given $\Psi \in F(c)$, say $\Psi : G \ltimes O \to C$, the associated Ψ^1 fits into the following commutative diagram:

$$
\begin{array}{ccc}
G \ltimes TO & \xrightarrow{T\Psi} & TC \\
\tau_O \downarrow & & \downarrow \tau \\
O^1 & \xrightarrow{\Psi^1} & \mathcal{C}^1
\end{array}
\tag{4.4}
$$

Here we already have an example of an sc-smooth functor. Namely, $\tau : TC \to \mathcal{C}^1$ has for every $T\Psi$ an sc-smooth representative, namely, τ_O. One of the main points of having polyfold structures is to say that certain functors defining algebraic structures are sc-smooth, or that they are sc-Fredholm functors, which allows a perturbation theory. Also, if $\mathcal{C}$ is equipped with a polyfold structure, we can define sc-differential forms as certain kind of functors.

We can define the k-fold product category $TC \times .. \times TC$, which projects to a k-fold product of $\mathcal{C}^1$ with itself and pull back by the multi-diagonal, which we denote by

$$\oplus_{i=1}^k TC \to \mathcal{C}^1.$$

The preimages of objects are k-fold products of Banach spaces. Viewing $\mathbb{R}$ as a category with only the identities as morphisms, we are interested in certain functors

$$\omega : \oplus_{i=1}^k TC \to \mathbb{R}$$

that are multilinear and skew-symmetric on the fibers. We assume TC equipped with its canonical polyfold structure TF. Given $T\Psi \in \bar{F}(c)$, we can pull back ω via $\oplus_{i=1}^k T\Psi$

and obtain $\omega_\Psi : TO \oplus \cdots \oplus TO \to \mathbb{R}$. In [31], we have introduced the notion of an sc-smooth differential form.

Definition 4.99. *The functor ω is said to be sc-smooth provided there exist a family $(\Psi_\lambda)_{\lambda \in \Lambda}$ (Λ a set) of good uniformizers associated to F such that the collection of sets $|\Psi_\lambda(O_\lambda)|$ covers $|\mathcal{C}|$, and in addition all the ω_{Ψ_λ} are sc-smooth differential forms.*

Several remarks are in order.

Remark 4.100.

1. One can show that the definition does not depend on the choice of the family (Ψ_λ).

2. We view ω as associated to $\mathcal{C}$, despite the fact that it is defined on $T\mathcal{C}$, which lies over $\mathcal{C}^1$ and therefore seemingly only involves $\mathcal{C}^1$. However, using the good uniformizers for $\mathcal{C}$ in the definition of $T\mathcal{C}$ incorporates the polyfold structure on $\mathcal{C}$ in a subtle way.

3. One is tempted to call ω as defined above an sc-smooth differential form on $\mathcal{C}$. If we call ω according to (2) as associated to $\mathcal{C}$, it will turn out that $d\omega$, the exterior differentiation, can only be defined as a form associated to $\mathcal{C}^1$. Since we have the system of inclusion functors $\cdots \to \mathcal{C}^{i+1} \to \mathcal{C}^i \to \cdots \to \mathcal{C}$, we can take a direct limit for forms. The set (see below) of all $[\omega]$ defined by the direct limit for all $k \geq 0$ turns out to be invariant under d defined by $d[\omega] = [d\omega]$, so that it is better to call $[\omega]$ an sc-smooth differential form. Using this, we obtain a de Rham complex associated to $\mathcal{C}$, as we shall see below. For the moment, we shall call ω an sc-differential form, and leave the name sc-smooth differential form for the result of a further construction.

The collection of all sc-smooth functors ω is a set, since it is completely determined by the set (ω_{Ψ_λ}). Using the inclusion functors $\mathcal{C}^{i+1} \to \mathcal{C}^i$, we can pull back a functor ω defined on $\oplus_{i=1}^k T(\mathcal{C}^i)$ to $\oplus_{i=1}^k T(\mathcal{C}^{i+1})$. This is nothing else but restricting ω to tangent vectors of specified higher regularity. Denote by $\Omega^k(\mathcal{C}^i)$ the set of differential k-forms on $\mathcal{C}^i$, i.e., defined on $\oplus_{j=1}^k T\mathcal{C}^i$. This set has the obvious structure as a real vector space. Then we have the direct system

$$\to \Omega^k(\mathcal{C}^i) \to \Omega^k(\mathcal{C}^{i+1}) \to \cdots$$

and we denote the direct limit by $\Omega_\infty^k(\mathcal{C})$ and its elements by $[\omega]$. If (X, β) is an ep-groupoid constructed from the polyfold structure F on $\mathcal{C}$, we can use the equivalence $\beta : X \to \mathcal{C}$ to pull back $[\omega]$, since $T\beta : TX \to T\mathcal{C}$ is well defined. In fact, this pullback completely determines $[\omega]$ and is compatible with the exterior derivative defined on X; see [28, 31, 32]. As it turns out, the exterior differential is well defined, so that we obtain the de Rham complex

$$(\Omega_\infty^*(\mathcal{C}, F), d).$$

So, in particular, there exists a de Rham cohomology. We refer the reader for details of this theory to [28, 31, 32].

4.4.3 Finite-dimensional, branched, weighted suborbifolds

Suppose that $\mathcal{C}$ is a GCT equipped with a polyfold structure F. We are interested in certain full subcategories that will arise when studying Fredholm functors later on. In a first step, consider the non-negative rational numbers $\mathbb{Q}^+$ as objects in a category with the morphisms being the identities. Of interest for us are certain functors

$$\Theta : \mathcal{C} \to \mathbb{Q}^+.$$

In order to define this class of functors, we need the definition of a submanifold M of an M-polyfold X.

Definition 4.101. *Let X be an M-polyfold and M a subset. We say M is a submanifold provided for every $m \in M$ there exists an open neighborhood $U = U(m) \subset X$ and an sc-smooth map $r : U \to U$ having the following properties:*

(i) $r(U_i) \subset U_{i+1}$ for all i and $r : U \to U^1$ is sc-smooth.

(ii) $r \circ r = r$.

(iii) $r(U) = U \cap M$.

First of all, we note that r is an sc-smooth retraction, so that M is a subpolyfold. But the stronger requirement that $r : U \to U^1$ is sc-smooth, in fact, implies that the M-polyfold structure on M induced from X is sc-smoothly equivalent to the structure of a finite-dimensional smooth manifold; see [31]. This, of course, justifies our calling M a submanifold in the first place. More precisely we have the following result.

Proposition 4.102. *Let M be a submanifold of the M-polyfold X in the sense of Definition 4.101. Then M is a subpolyfold and its sc-smooth structure induced from X is sc-smoothly equivalent to a classical manifold structure.*

Remark 4.103. A hint that this is true is given by the following consideration. If $r(x) = x$, then it follows that $x \in X_\infty$, so that for every $m \in M$, the tangent space $T_m M$ is defined and is an sc-Banach space. Since $Tr(m) : T_m X \to T_m X$ is an sc^+-operator, its image is compact. Since $Tr(m)$ restricted to its image is the identity, the image must be finite-dimensional.

If M, viewed as manifold, is equipped with an orientation, we shall call it an oriented submanifold.

Definition 4.104. *Suppose $\Theta : \mathcal{C} \to \mathbb{Q}^+$ is a functor and c an object. Pick $\Psi \in F(c)$, so we can take the functor $\Theta \circ \Psi : G \ltimes O \to \mathbb{Q}^+$. Let $\Psi(q_0) = c$ and assume that there exist an open neighborhood $U(q_0)$ in O, finitely many submanifolds $(M_i)_{i \in I}$ and positive rational numbers $(\sigma_i)_{i \in I}$ such that*

$$\Theta \circ \Psi(q) = \sum_{\{i \in I \,\mid\, q \in M_i\}} \sigma_i$$

for all $q \in U(q_0)$. We say that Θ has a smooth finite-dimensional representation with respect to Ψ at c. We also say that the representation is n-dimensional provided every M_i is n-dimensional.

If $\Psi' \in F(c)$ with $\Psi'(q_0') = c$, we consider $\mathbf{M}(\Psi, \Psi')$ and find open neighborhoods such that the source and target maps are sc-diffeomorphisms

$$V(q_0) \xleftarrow{s} V(q_0, 1_c, q_0') \xrightarrow{t} V(q_0').$$

We may assume that $V(q_0) \subset U(q_0)$ and that we can map the $M_i \cap V(q_0)$ to $M_i' \subset V(q_0')$. Then

$$\Theta \circ \Psi'(q') = \sum_{\{i \in I \,\mid\, q \in M_i'\}} \sigma_i$$

for $q' \in V(q_0')$. Hence if we have a smooth finite-dimensional representation at c for some $\Psi \in F(c)$, then it holds for all uniformizers in $F(c)$. The same argument goes through if $\phi : c \to c'$ is an isomorphism and shows that if we have a smooth finite-dimensional representation at c, we also have this property at an isomorphic c'. Observe that if the representation at c with respect to Ψ is n-dimensional, this will be true for every isomorphic c as well. We can therefore say that Λ is smooth at $|c|$ and has a n-dimensional representation at $|c|$. Since the collection of all $|c|$ is a set, the following makes sense.

Definition 4.105. *Let C be a GCT equipped with a polyfold structure and let $\Theta : C \to \mathbb{Q}^+$ be a functor. We call Θ a smooth, weighted, branched suborbifold of dimension n provided that Θ has a smooth n-dimensional representation at every $|c| \in |C|$.*

There is also a notion of orientation for Θ; see [28, 29, 32].

Definition 4.106. *Let C be a polyfold category and $\Theta : C \to \mathbb{Q}^+$ a smooth, weighted, branched suborbifold of dimension n. We say Θ is closed provided that the orbit space associated to all objects c with $\Theta(c) > 0$ is a compact subset of $|C|$.*

An important result is that we can integrate sc-differential forms over closed, smooth, oriented, weighted branched suborbifolds of dimension n; see [28, 29, 32].

Theorem 4.107. *Let C be a GCT equipped with a polyfold structure F and let $\Theta : C \to \mathbb{Q}^+$ be a closed, smooth, oriented, weighted, branched suborbifold of dimension n. Suppose further that $[\omega]$ is an n-dimensional sc-smooth differential form on C. Then there is a well-defined integral*

$$\oint_\Theta [\omega],$$

called the branched integral.

The integral is characterized uniquely by certain properties, and in the somewhat more general context with a boundary with corners, even a version of Stokes' theorem holds. The basic observation is that Θ defines a compact subset of $|\mathcal{C}|$ equipped with what is called a Lebesgue σ-algebra of measurable sets and a measurable $\mathbb{Q}^+$-valued weight function w. The differential form $[\omega]$ induces a signed measure $\mu_{[\omega]}$, and the integral is given by $\int w\, d\mu_{[\omega]}$. This construction is formidable and is given in [28, 29] for ep-groupoids, but generalizes immediately to our context, since the ep-groupoid version is compatible with Morita equivalence, i.e., generalized isomorphisms associated to diagrams of sc-smooth equivalences between ep-groupoids; see [32].

4.4.4 Fredholm theory and transversality

Suppose we are given a bundle GCT $(\mathcal{C}, \mu, \mathcal{T})$ equipped with a strong polyfold bundle structure $\bar{F} : \mathcal{C} \to$ SET and a section functor

$$f : \mathcal{C} \to \mathcal{E}_\mu.$$

If c is an object in $\mathcal{C}$ and $\bar{\Psi}$ a strong bundle uniformizer around c, then we obtain the commutative diagram.

$$
\begin{array}{ccc}
G \ltimes K & \xrightarrow{\;\bar{\Psi}\;} & \mathcal{E}_\mu \\
{\scriptstyle p}\downarrow & & \downarrow{\scriptstyle P_\mu} \\
G \ltimes O & \xrightarrow{\;\Psi\;} & \mathcal{C}
\end{array}
$$

Since f maps an object to an element in the Banach space associated to c, i.e., $P^{-1}(c)$, it follows that an object $q \in O$ is mapped to $f \circ \Psi(q) \in P^{-1}(\Psi(q))$. However, $\bar{\Psi} : p^{-1}(q) \to P^{-1}(\Psi(q))$ is a linear isomorphism and it follows that f has a local representative f_Ψ that is a section of p.

Definition 4.108. *Let f be a section functor of $P : \mathcal{E}_\mu \to \mathcal{C}$, where $(\mathcal{C}, \mu, \mathcal{T})$ is a bundle GCT equipped with a strong polyfold bundle structure $\bar{F}$. We say that f is an sc-Fredholm functor provided there exists a family $(\bar{\Psi}_\lambda)_{\lambda \in \Lambda}$ (Λ a set) of good strong bundle uniformizers such that $(\bar{\Psi}_\lambda)_{\lambda \in \Lambda}$ covers $|\mathcal{C}|$ and every $f_{\bar{\Psi}_\lambda}$ is an sc-Fredholm section of $K_\lambda \to O_\lambda$.*

This definition does not depend on the choice of the strong bundle uniformizers taken from $\bar{F}$.

Of course, as in the classical situation, one is interested in a perturbation theory, which allows an sc-Fredholm section to be brought into a general position. Since we are dealing with functors, there is the added difficulty that symmetry, i.e., compatibility with morphisms, and transversality are competing issues. In order to achieve transversality, we locally break the symmetry under a small perturbation by an sc$^+$-section, but keep track of the symmetry by introducing a finite family (correlated with the initial perturbation) of local sc-Fredholm problems invariant under the symmetry. In order to still have the right counts of the solutions, the problems in the local family need to be weighted. Of course, these local modifications have to be done coherently, so that overlapping families can be patched together in a suitable way. In this context, it is very important to understand how big perturbations can be, in order to guarantee that the perturbed Fredholm section is again proper. In order to formulate some results, we need some auxiliary structures. We view $\mathbb{R}^+ \cup \{+\infty\} = [0, +\infty]$ as a category only having the identities as morphisms.

Definition 4.109. *Let $(\mathcal{C}, \mu, \mathcal{T})$ be a bundle GCT equipped with a strong polyfold bundle structure $\bar{F} : \mathcal{C} \to \text{SET}$. An auxiliary norm N is a functor $N : \mathcal{E}_\mu \to \mathbb{R}^+ \cup \{+\infty\}$ with the following properties. There exists a family of good strong bundle uniformizers $(\bar{\Psi}_\lambda)_{\lambda \in \Lambda}$ (Λ a set) such that the underlying $(|\Psi_\lambda(O_\lambda)|)_{\lambda \in \Lambda}$ cover $|\mathcal{C}|$ and for every $\lambda \in \Lambda$,*

$$N \circ \bar{\Psi}_\lambda : K_\lambda \to \mathbb{R}^+ \cup \{+\infty\}$$

is an auxiliary norm according to Definition 4.87.

The definition does not depend on the family of good strong bundle uniformizers.

Definition 4.110. *Let $(\mathcal{C}, \mu, \mathcal{T})$ be a bundle GCT equipped with a strong polyfold bundle structure $\bar{F} : \mathcal{C} \to \text{SET}$. A functor $\Lambda : \mathcal{E}_\mu \to \mathbb{Q}^+$ is called an sc$^+$-smooth multisection functor provided that there exists a set $(\bar{\Psi}_\lambda)_{\lambda \in A}$ of good strong bundle uniformizers with the underlying (Ψ_λ) satisfying*

$$|\mathcal{C}| = \bigcup_{\lambda \in A} |\Psi_\lambda(O_\lambda)|$$

and with the following property:

- *For every $\Lambda \circ \bar{\Psi}_\lambda : G_\lambda \ltimes K_\lambda \to \mathbb{Q}^+$, and given $q \in O_\lambda$, there exist an open neighborhood $U(q) \subset O_\lambda$, sc$^+$-sections $(s_i)_{i \in I}$ defined for $K_\lambda | U(q)$, and rational weights $\sigma_i > 0$ ($i \in I$) with $\sum_{i \in I} \sigma_i = 1$, such that for $e \in K_\lambda | U(q)$,*

$$\Lambda \circ \bar{\Psi}_\lambda(e) = \sum_{\{i \in I \mid e = s_i(p_\lambda(e))\}} \sigma_i.$$

Remark 4.111. The definition is independent of the choice of $(\bar{\Psi}_\lambda)$ as long as the associated open sets $(\Psi(O_\lambda))$ cover the orbit space of $\mathcal{S}^{3,\delta_0}(Q, \omega)$. The functor Λ induces a map $|\Lambda| : |\mathcal{E}_\mu| \to \mathbb{Q}^+$.

Definition 4.112. *The support of an sc$^+$-smooth multisection functor Λ is the full subcategory $\text{supp}(\Lambda)$ associated to all objects e in $\mathcal{E}_\mu$ with $\Lambda(e) > 0$. The domain support dom-$\text{supp}(\Lambda)$ is the full subcategory associated to the closure of the open subset U of $|\mathcal{C}|$ consisting of all $|c|$ such that there exists $|e| \in |P|^{-1}(|c|)$, $|e| \neq 0$, with $|\Lambda|(|e|) > 0$.*

We note that each fiber of $|P|$ has a distinguished 0 element. Given $\Lambda : \mathcal{E}_\mu \to \mathbb{Q}^+$ and an auxiliary norm $N : \mathcal{E}_\mu \to [0, \infty]$, we can measure the size of Λ. Namely, for an object c, there are finitely many points $e_i \in P_\mu^{-1}(c)$ for which $\Lambda(e_i) > 0$.

Moreover, the bi-regularity is at least $(0, 1)$, so $N(e_i)$ is finite. Therefore, we can define $\max\{N(e_i) \mid i \in I\}$ and obtain a functor $\mathcal{C}_\mu \to [0, \infty)$ by

$$c \to \max\{N(e_i) \mid i \in I\}.$$

We can pass to orbit space and obtain a continuous map $N_\Lambda : |\mathcal{C}| \to [0, \infty)$.

Now we are in a position to describe a series of results in sc-Fredholm theory for our categorical setup. Without going into much detail, there is a notion of orientation for an sc-smooth Fredholm section functor; see [30–32]. The first result is concerned with a compactness assertion.

Theorem 4.113. *Let $(\mathcal{C}, \mu, \mathcal{T})$ be a bundle GCT equipped with a strong polyfold bundle structure $\bar{F} : \mathcal{C} \to \text{SET}$. Suppose that $N : \mathcal{E}_\mu \to \mathbb{R}^+ \cup \{+\infty\}$ is an auxiliary norm and f an sc-Fredholm section functor of P_μ, having the property that the orbit space $|f^{-1}(0)| \subset |\mathcal{C}|$ is compact. Then there exists an open neighborhood U of $|f^{-1}(0)|$ in $|\mathcal{C}|$ such that for every sc^+-multisection functor $\Lambda : \mathcal{E}_\mu \to \mathbb{Q}^+$ with domain support in $\mathcal{C}_U$ and satisfying $N_\Lambda(|c|) \leq 1$ for all $|c| \in |\mathcal{C}|$, the orbit space associated to $\text{supp}(\Lambda \circ f)$ is compact.*

Note that $\Lambda \circ f : \mathcal{C} \to \mathbb{Q}^+$ is a functor and $\text{supp}(\Lambda \circ f)$ consists of all objects c with $\Lambda \circ f(c) > 0$. Moreover, the functor $\Lambda \circ f$ takes its values in $[0, 1] \cap \mathbb{Q}^+$.

Definition 4.114. *If (U, N) is a pair consisting of an auxiliary norm N and an open neighborhood U of the orbit space associated to $f^{-1}(0)$ such that the conclusion of Theorem 4.113 concerning compactness holds, we shall say that (U, N) controls compactness.*

In order to construct sc^+-multisection functors, one needs the underlying polyfold structure on $\mathcal{C}$ to admit sc-smooth partitions of unity. For example if everything is built on Hilbert scales, or at least the zero-level, then these are available; see [31, 32]. In the following, we denote by $H^*_{dR}(\mathcal{C})$ the de Rham cohomology.

Theorem 4.115. *Let $(\mathcal{C}, \mu, \mathcal{T})$ be a bundle GCT equipped with a strong polyfold bundle structure $\bar{F} : \mathcal{C} \to \text{SET}$. Suppose that $N : \mathcal{E}_\mu \to \mathbb{R}^+ \cup \{+\infty\}$ is an auxiliary norm and f an sc-Fredholm section functor of P_μ having the property that the orbit space $|f^{-1}(0)| \subset |\mathcal{C}|$ is compact. We assume that the induced polyfold structure on $\mathcal{C}$ admits sc-smooth partitions of unity. Let U be an open neighborhood around the orbit space associated to $f^{-1}(0)$ such that (U, N) controls compactness. Then the following hold:*

(i) *Given any $\varepsilon \in (0, 1)$, there exists an sc^+-multisection functor Λ with domain support in $\mathcal{C}_U$ and $N_\Lambda(|c|) < \varepsilon$ for all $|c|$ such that that*

$$\Theta := \Lambda \circ f : \mathcal{C} \to \mathbb{Q}^+$$

is a smooth, closed, weighted, branched suborbifold of dimension n.

(ii) *If f is oriented, i.e., $(f, \mathfrak{o})$, then Θ is naturally oriented.*

(iv) *Given $[\omega] \in H^n_{dR}(\mathcal{C})$, the branched integral $\oint_\Theta [\omega]$ does not depend on the choice of Λ as long as it is generic and satisfies the above conditions.*

(v) *The value of the integral is independent of the choice (U, N) as long as it is admissible.*

As a consequence of this theorem, the oriented sc-Fredholm section functor (f, o) defines a linear functional

$$I_{(f,o)} : H^*_{dR}(\mathcal{C}) \to \mathbb{R}$$

via

$$I_{(f,o)}([[\omega]]) = \oint_{\Lambda \circ f} [\omega],$$

for Λ that has sufficiently small support and is generic. This functional will stay the same under even large oriented deformations

$$t \to (f_t, o_t)$$

as long as the orbit space of the solution set satisfies some compactness properties. During the deformation, we can even change the μ_t as long as this is done sc-smoothly for an overall strong polyfold bundle structure. There are also appropriate versions where the underlying category changes (sc-smoothly) as well.

4.4.5 The stable map example

In some sense, we just need to take a fresh look at what we did in Section 4.2 and verify, implementing the discussion from Section 4.3, that the constructions are sc-smooth. We shall concentrate on the category $\mathcal{S}^{3,\delta_0}(Q, \omega)$ rather than the full problem $P : \mathcal{E}^{3,\delta_0}(Q, \omega, J) \to \mathcal{S}^{3,\delta_0}(Q, \omega)$. In the case of $\mathcal{S}^{3,\delta_0}(Q, \omega)$, our discussion so far gives a precise construction with established topological properties. From the discussion in [30], based on results in [33], it follows that indeed all the constructions are sc-smooth.

Remark 4.116. It is a very good exercise to use [33] and some of the results in [30] to fill in the technical details.

We make a few useful comments. In the constructions mentioned in Theorem 4.51, we already used the exponential gluing profile. Given the strictly increasing sequence $\delta = (\delta_0, \delta_1, \dots)$ with all $0 < \delta_i < 2\pi$, it can be shown that $\bar{X}^{3,\delta_0}_{(S,j,D),D,\mathcal{H}}(Q)$ has an M-polyfold structure. It only depends on δ and we shall abbreviate this topological space equipped with this M-polyfold structure by $\bar{X}^{3,\delta}_{(S,j,D),D,\mathcal{H}}(Q)$. As a consequence, the product

$$V \times \bar{X}^{3,\delta}_{(S,j,D),D,\mathcal{H}}(Q)$$

also has an M-polyfold structure. For this structure, the automorphism group G acts by sc-diffeomorphisms. Taking a suitable open G-invariant neighborhood O of $(0, (S, D, u))$, we obtain

$$\Psi : G \ltimes O \to \mathcal{S}^{3,\delta_0}(Q, \omega), \tag{4.5}$$

which on objects is given by

$$(v, (S_a, D_a, w)) \rightarrow (S_a, j(v)_a, M_a, D_a, w).$$

Also recall that we had the underlying good uniformizers for $\mathcal{R}$,

$$G \ltimes O^* \rightarrow \mathcal{R},$$

which on objects map

$$(v, a) \rightarrow (S_a, j(v)_a, M_a^*, D_a).$$

Our aim is to define a polyfold structure on $\mathcal{S}^{3,\delta_0}(Q, \omega)$, utilizing the previous construction of F. Of course, we need to be able to equip $\mathbf{M}(\Psi, \Psi')$ with an M-polyfold structure. In order to achieve this, we have to impose an additional requirement, which is achieved by possibly restricting Ψ to a smaller O. This leads to the definition of a good (polyfold) uniformizer.

Definition 4.117. *Assume that a sequence δ and the exponential gluing profile are given as previously described. Let $\alpha = (S, j, M, D, u)$ be an object in $\mathcal{S}^{3,\delta_0}(Q, \omega)$ with automorphism group G. A good polyfold uniformizer α for $\mathcal{S}^{3,\delta_0}(Q, \omega)$ is a functor $\Psi : G \ltimes O \rightarrow \mathcal{S}^{3,\delta_0}(Q, \omega)$ as constructed in (4.5), where O is a G-invariant open neighborhood of $(0, (S, D, u))$ in $V \times \bar{X}^{3,\delta}_{(S,j,D),D,\mathcal{H}}(Q)$ such that the following hold:*

(i) Ψ is fully faithful and $\Psi(v, (S, D, u)) = (S, j, M, D, u)$.

(ii) Passing to orbit spaces, $|\Psi| : {}_G\backslash O \rightarrow |\mathcal{S}^{3,\delta_0}(Q, \omega)|$ is a homeomorphism onto an open neighborhood of $|\alpha|$.

(iii) The collection of all (v, a) occurring in elements $(v, (S_a, D_a, w)) \in O$ are contained in O^*, and $O^* \ni (v, a) \rightarrow (S_a, j(v)_a, M_a^*, D_a)$ defines a good uniformizer for $\mathcal{R}$.

Remark 4.118.

(a) The condition (iii) is very important and it summarizes four conditions from the definition of a good uniformizer for $\mathcal{R}$. Most important is that the partial Kodaira–Spencer differentials are isomorphisms. This is extensively used when putting an M-polyfold structure on $\mathbf{M}(\Psi, \Psi')$. It is crucial for making s and t local sc-diffeomorphisms—see [30].

(b) Let $F(\alpha)$ consist of the good uniformizers previously constructed and their domain $G \ltimes O$ equipped with the M-polyfold structures associated to a choice of δ. In addition, we assume that the set of all (v, a) coming from the $(v, (S_a, D_a, w))$ lies in O^*, so that we have the good uniformizer for $\mathcal{R}$ defined on $G \ltimes O^*$. With F modified as just described it, is possible to lift $\mathbf{M}$ to a functor that associates to (Ψ, Ψ') not only a metrizable space, but in fact a n M-polyfold structure. The details given in the ep-groupoid seen are presented in [30].

The relevant construction

$$F : \mathcal{S}^{3,\delta_0}(Q, \omega) \to \text{SET}$$

associates to an object α the set $F(\alpha)$ consisting of good polyfold uniformizers as described in Definition 4.117. This leads to the following important result.

Theorem 4.119. *Given the exponential gluing profile φ and an increasing sequence of weights $0 < \delta_0 < \delta_1 < \cdots < 2\pi$, the construction $(F, \mathbf{M})$ given in Theorem 4.51, with the modification just mentioned, defines a polyfold structure, when the local models are equipped with the sc-structures associated to these weights.*

We can also construct a bundle GCT $(\mathcal{S}^{3,\delta_0}(Q, \omega), \mu, \mathcal{T})$ as follows. Given an object $\alpha = (S, j, M, D, u)$, we associate to it the Hilbert space $\mu(\alpha)$ consisting of all (TQ, J)-valued $(0, 1)$-forms ξ along u of class $(2, \delta_0)$. In particular, $\xi(z) : (T_z S, j) \to (T_{u(z)} Q, J)$ is complex antilinear. A morphism $\Phi := (\alpha, \phi, \alpha') : \alpha \to \alpha'$ defines a linear topological isomorphism

$$\mu(\Phi) : \mu(\alpha) \to \mu(\alpha') : \xi \to \xi \circ T\phi^{-1}.$$

This allows us to define the category $\mathcal{E}_\mu$ whose objects are the (α, ξ), where α is an object in $\mathcal{S}^{3,\delta_0}(Q, \omega)$ and ξ is a vector in $\mu(\alpha)$. We shall denote $\mathcal{E}_\mu$ by $\mathcal{E}^{2,\delta_0}(Q, \omega, J)$. One can define a metrizable topology $\mathcal{T}$ for $|\mathcal{E}^{2,\delta_0}(Q, \omega, J)|$ and carry out a construction $(\bar{F}, \mathbf{M})$,

$$\bar{F} : \mathcal{S}^{3,\delta_0}(Q, \omega) \to \text{SET},$$

equipping P with a strong polyfold bundle structure. Here $\bar{F}(\alpha)$ consists of good strong bundle uniformizers, where an element $\bar{\Psi}$ in the latter fits into a commutative diagram of functors with certain properties:

$$
\begin{array}{ccc}
G \ltimes K & \xrightarrow{\bar{\Psi}} & \mathcal{E}^{2,\delta_0}(Q, \omega, J) \\
{\scriptstyle p}\downarrow & & {\scriptstyle p}\downarrow \\
G \ltimes O & \xrightarrow{\Psi} & \mathcal{S}^{3,\delta_0}(Q, \omega)
\end{array}
$$

On objects, Ψ has the form

$$\Psi(v, (S_a, D_a, w)) = (S_a, j(v)_a, M_a, w)$$

and $\bar{\Psi}$ is given by

$$\bar{\Psi}((v, (S_a, D_a, w), \xi) = (S_a, j(v)_a, M_a, w, \xi).$$

The transition structure

$$\mathbf{M}(\bar{\Psi}, \bar{\Psi}') \to \mathbf{M}(\Psi, \Psi')$$

has an sc-smooth strong bundle structure. These structures, as already explained, can be viewed as some kind of sc-smooth bundle structure for $P : \mathcal{E}^{2,\delta_0}(Q, \omega, J) \to S^{3,\delta_0}(Q, \omega)$. The latter, equipped with this strong bundle structure, which depends on the weight sequence δ, is written as

$$P : \mathcal{E}^{2,\delta}(Q, \omega, J) \to S^{3,\delta}(Q, \omega).$$

The local representative of the Cauchy–Riemann section takes the form

$O \to K :$

$$(v, (S_a, D_a, w)) \to \left(v, (S_a, D_a, w), \tfrac{1}{2}\left[Tw + J(w) \circ Tw \circ j(v)_a\right]\right). \quad (4.6)$$

It has been proved in [30] that the section just defined is sc-Fredholm.

Theorem 4.120. *The local representative of the section functor $\bar{\partial}_J$ of P with respect to $\bar{\Psi} \in \bar{F}(\alpha)$ for any object α, as given in (4.6), is sc-Fredholm. Hence $\bar{\partial}_J$ is an sc-Fredholm section of $P : \mathcal{E}^{2,\delta}(Q, \omega, J) \to S^{3,\delta}(Q, \omega)$.*

In the case of our Gromov–Witten example, consider the full subcategory $S^{3,\delta}_{g,m,A}(Q, \omega)$ of $S^{3,\delta}(Q, \omega)$, where $g \geq 0$ and $m \geq 0$ are integers, and $A \in H_2(Q, \mathbb{Z})$, consisting of all stable maps of arithmetic genus g with m marked points in the homology class A.

Theorem 4.121. *The following hold:*

(i) *The orbit space $|S^{3,\delta}_{g,m,A}(Q, \omega)|$ is an open and closed subset of $|S^{3,\delta}(Q, \omega)|$ and consequently $S^{3,\delta}_{g,m,A}(Q, \omega)$ has an induced polyfold structure. Moreover, $\mathcal{E}^{2,\delta}_{g,m,A}(Q, \omega, J) := \mathcal{E}^{2,\delta}(Q, \omega, J)|S^{3,\delta}_{g,m,A}(Q, \omega)$ has an induced strong polyfold bundle structure.*

(ii) *The orbit space of $(f|S^{3,\delta}_{g,m,A}(Q, \omega))^{-1}(0)$ is compact.*

(iii) *For every $1 \leq i \leq m$, the evaluation map*

$$ev_i : S^{3,\delta}_{g,m,A}(Q, \omega) \to Q \quad (4.7)$$

at the ith marked point is an sc-smooth functor in the sense that $ev_i \circ \Psi$ is sc-smooth if Ψ is a good uniformizer.

(iv) *If $2g + m \geq 3$, then the forgetful functor*

$$\sigma : S^{3,\delta}_{g,m,A}(Q, \omega) \to \mathcal{R}^{ord}_{g,m} \quad (4.8)$$

into the stable Riemann surface category with ordered marked points is sc-smooth as well, where we can take good uniformizers for $S^{3,\delta}_{g,m,A}(Q, \omega)$ and $\mathcal{R}^{ord}_{g,m}$ to see that the local representative is sc-smooth.

The consequence of (iii) is that the pullback of a smooth differential form on Q gives an sc-smooth differential form on $S^{3,\delta}_{g,m,A}(Q, \omega)$. For the forgetful map, the pullback of a smooth differential form defines an sc-smooth differential form on $\mathcal{S}^{3,\delta}_{g,m,A}(Q,\omega)$ as a consequence of (iv).

Moreover, $\bar{\partial}_J$ restricted to $\mathcal{S}^{3,\delta}_{g,m,A}(Q,\omega)$ is an sc-smooth Fredholm functor with a compact solution set, and it has a natural orientation, giving us $(\bar{\partial}_J, \mathfrak{o})$ and its restrictions $(\bar{\partial}_{J,(g,k,A)}, \mathfrak{o})$. As a consequence, we have the linear maps $I_{(g,m,A)} := I_{(\bar{\partial}_{J,(g,k,A)},\mathfrak{o})}$:

$$I_{(g,m,A)} : H^*_{dR}(\mathcal{S}^{3,\delta}_{g,k,A}(Q,\omega)) \to \mathbb{R}.$$

These maps give precisely the data from which one can define the Gromov–Witten invariants—see [30].

Remark 4.122. There is, of course, a large literature on Gromov–Witten invariants and further developments—see [14, 15, 38, 39, 42–45, 52, 54, 55, 58]. All the methods differ. In some sense, all the approaches had to come up with a fix for the fact that classical Fredholm theory doesn't work.

The theory that we have described here provides a very powerful language to deal with moduli problems in symplectic geometry. In [12], we use this approach to define a Fredholm setup for SFT. The language for doing so is that developed in [31, 32]. The necessary nonlinear analysis comes from [33].

Acknowledgments

The author thanks the Clay Institute for the opportunity to present this material at the 2008 Clay Research Conference. Some of the initial work was done during a sabbatical at the Stanford Mathematics Department and was supported in part by the American Institute of Mathematics. Thanks to P. Albers, Y. Eliashberg, J. Fish, E. Ionel, K. Wehrheim, K. Wysocki and E. Zehnder for many stimulating discussions.

References

[1] A. Adem, J. Leida and Y. Ruan, *Orbifolds and Stringy Topology* (Cambridge University Press, 2007).

[2] V. Borisovich, V. Zvyagin and V. Sapronov, *Nonlinear Fredholm maps and Leray–Schauder degree*, Russ. Math. Surv. **32**(4) (1977), 1–54.

[3] F. Bourgeois, Y. Eliashberg, H. Hofer, K. Wysocki and E. Zehnder, *Compactness results in symplectic field theory*, Geom. Topol. **7** (2003) 799–888.

[4] H. Cartan, *Sur les rétractions d'une variété*, C. R. Acad. Sci. Paris, Ser. I **303** (1986) 715–716.

[5] K. Cieliebak, I. Mundet i Riera and D. A. Salamon, *Equivariant moduli problems, branched manifolds, and the Euler class*, Topology **42** (2003) 641–700.

[6] P. Deligne and D. Mumford, *The irreducibility of the space of curves of given genus*, Inst. Hautes Études Sci. Publ. Math. **36** (1969) 75–109.

[7] S. Donaldson and P. Kronheimer, *The Geometry of Four-Manifolds* (Oxford University Press, 1990).

[8] D. Ebin, *The manifold of Riemannian metrics*, Proc. Symp. Pure Math. **15** (1968) 11–40.

[9] Y. Eliashberg, A. Givental and H. Hofer, *Introduction to symplectic field theory*, in N. Alon, J. Bourgain, A. Connes, M. Gromov and V. D. Milman, eds., *Visions in Mathematics: GAFA 2000 Special Volume*, Part II (Birkhäuser, 2000), pp. 560–673.

[10] H. Eliasson, *Geometry of manifolds of maps*, J. Differential Geom. **1**(1967) 169–194.

[11] O. Fabert, J. W. Fish, R. Golovko and K. Wehrheim, *Polyfolds: a first and second look*, arXiv:1210.6670 [math.SG].

[12] J. Fish and H. Hofer, *Polyfold constructions*, in preparation.

[13] A. Floer and H. Hofer, *Coherent orientations for periodic orbit problems in symplectic geometry*, Math. Z. **212** (1993) 13–38.

[14] K. Fukaya and K. Ono, *Arnold conjecture and Gromov–Witten invariants*, Topology **38** (1999) 933–1048.

[15] K. Fukaya, Y.-G. Oh, H. Ohta and K. Ono, *Lagrangian Intersection Floer Theory: Anomaly and Obstruction*, Part I (American Mathematical Society/International Press, 2009).

[16] K. Fukaya, Y.-G. Oh, H. Ohta and K. Ono, *Lagrangian Intersection Floer Theory: Anomaly and Obstruction*, Part II (American Mathematical Society/International Press, 2009).

[17] P. Gabriel and M Zisman, *Calculus of Fractions and Homotopy Theory* (Springer-Verlag, 1967).

[18] M. Gromov, *Pseudoholomorphic curves in symplectic geometry*, Inv. Math. **82** (1985) 307–347.

[19] A. Haefliger, *Homotopy and integrability*, in N. H. Kuiper, ed. *Manifolds (Amsterdam, 1970)*, Lecture Notes in Mathematics, Vol. 197 (Springer-Verlag, 1971), pp. 133–163.

[20] A. Haefliger, *Holonomie et classifiants*, Asterisque **116** (1984) 70–97.

[21] A. Haefliger, *Groupoids and foliations*, Contemp. Math. **282** (2001) 83–100.

[22] H. Hofer, *A general Fredholm theory and applications*, in D. Jerison, B. Mazur, T. Mrowka, W. Schmid, R. Stanley and S. T. Yau, eds., *Current Developments in Mathematics* (International Press, 2006), pp. 1–71.

[23] H. Hofer, *Polyfolds and a general Fredholm theory*, arXiv:0809.3753 [math.SG].

[24] H. Hofer, K. Wysocki and E. Zehnder, *Deligne–Mumford-type spaces with a view towards symplectic field theory*, lecture notes in preparation.

[25] H. Hofer, K. Wysocki and E. Zehnder, *A general Fredholm theory I: a splicing-based differential geometry*, J. Eur. Math. Soc. (JEMS) **9** (2007) 841–876.

[26] H. Hofer, K. Wysocki and E. Zehnder, *A general Fredholm theory II: implicit function theorems*, Geom. Funct. Anal. **19** (2009) 206–293.

[27] H. Hofer, K. Wysocki and E. Zehnder, *A general Fredholm theory III: Fredholm functors and polyfolds*, Geom. Topol. **13** (2009) 2279–2387.

[28] H. Hofer, K. Wysocki and E. Zehnder, *Integration theory for zero sets of polyfold Fredholm sections*, Math. Ann. **346** (2010) 139–198.

[29] H. Hofer, K. Wysocki and E. Zehnder, *Erratum: Integration Theory on the Zero Set of Polyfold Fredholm Sections*, preprint.

[30] H. Hofer, K. Wysocki and E. Zehnder, *Applications of polyfold theory I: the polyfolds of Gromov–Witten theory*, arXiv:1107.2097 [math.SG] (Mem. Am. Math. Soc., to appear).

[31] H. Hofer, K. Wysocki and E. Zehnder, *Polyfolds and Fredholm theory I: basic theory in M-polyfolds*, arXiv:1407.3185 [math.FA].

[32] H. Hofer, K. Wysocki and E. Zehnder, *Polyfold and Fredholm Theory, in preparation.*

[33] H. Hofer, K. Wysocki and E. Zehnder, *Sc-smoothness, retractions and new models for smooth spaces,* Discrete Continuous Dyn. Syst. **28** (2010) 665–788.

[34] M. Hutchings and J. Nelson, *Cylindrical contact homology for dynamically convex contact forms in three dimensions,* arXiv:1407.2898 [math.SG] (J. Symplectic Geom., to appear).

[35] D. Joyce, *Kuranishi bordism and Kuranishi homology,* arXiv:0707.3572 [math.SG].

[36] M. Kontsevich, *Enumeration of rational curves via torus action,* in R. Dijkgraaf, C. Faber and G. van der Geer, eds., *The Moduli Space of Curves* (Birkhäuser 1995), pp. 335–568.

[37] S. Lang, *Introduction to Differentiable Manifolds,* 2nd edn (Springer-Verlag, 2002).

[38] J. Li and G. Tian, *Virtual moduli cycles and Gromov–Witten invariants of general symplectic manifolds,* in R. J. Stern, ed., *Topics in Symplectic 4-Manifolds (Irvine, CA, 1996)* (International Press, 1998), pp. 47–83.

[39] G. Lu and G. Tian, *Constructing virtual Euler cycles and classes,* Int. Math. Res Surv. **2007** (2008) doi: 10.1093/imrsur/rym001.

[40] Y. Manin, *Frobenius Manifolds, Quantum Cohomology, and Moduli Spaces* (American Mathematical Society, 1999).

[41] D. McDuff, *Groupoids, branched manifolds and multisection,* J. Symplectic Geom. **4** (2006) 259–315.

[42] D. McDuff, *Notes on Kuranishi atlases,* arXiv:1411.4306 [math.SG].

[43] D. McDuff and K. Wehrheim, *The fundamental class of smooth Kuranishi atlases with trivial isotropy,* arXiv:1508.01560 [math.SG].

[44] D. McDuff and K. Wehrheim, *Smooth Kuranishi atlases with isotropy,* arXiv:1508.01556 [math.SG].

[45] D. McDuff and K. Wehrheim, *The topology of Kuranishi atlases,* arXiv:1508.01844 [math.SG].

[46] D. McDuff and D. Salamon, *Introduction to Symplectic Topology,* 2nd edn (Oxford University Press, 1998).

[47] D. McDuff and D. Salamon, *J-Holomorphic Curves and Symplectic Topology* (American Mathematical Society, 2004).

[48] I. Moerdijk, *Orbifolds as groupoids: an introduction,* Contemp. Math. **310** (2001) 205–222.

[49] I. Moerdijk and J. Mrčun, *Introduction to Foliations and Lie Groupoids* (Cambridge University Press, 2003).

[50] J. Nelson, *Automatic transversality in contact homology I: regularity,* arXiv:1407.3993 [math.SG].

[51] H. Omori, *Infinite Dimensional Lie Transformation Groups,* Lecture Notes in Mathematics, Vol. 427 (Springer-Verlag, 1974).

[52] J. Pardon, *An algebraic approach to virtual fundamental cycles on moduli spaces of pseudo-holomorphic curves,* Geom. Topol. **20** (2016) 779–1034.

[53] J. Robbin and D. Salamon, *A construction of the Deligne–Mumford orbifold,* J. Eur. Math. Soc. (JEMS) **8** (2006) 611–699.

[54] Y. Ruan, *Topological sigma model and Donaldson-type invariants in Gromov theory,* Duke Math. J. **83** (1996) 451–500.

[55] Y. Ruan, *Symplectic topology on algebraic 3-folds*, J. Differential Geom. **39** (1994) 215–227.

[56] S. Smale, *An infinite dimensional version of Sard's theorem*, Am. J. Math. **87** (1965) 861–866.

[57] S. Suhr and K. Zehmisch, *Polyfolds, cobordisms, and the strong Weinstein conjecture*, arXiv:1411.5016 [math.SG].

[58] G. Tian, *Quantum cohomology and its associativity*, in R. Bott, M. Hopkins, A. Jaffe, I. Singer, D. W. Stroock and S.-T. Yau, eds., *Current Developments in Mathematics* (International Press, 1995), pp. 361–397.

[59] H. Triebel, *Interpolation Theory, Function Spaces, Differential Operators* (North-Holland, 1978).

[60] C. Wendl, *Automatic transversality and orbifolds of punctured holomorphic curves in dimension four*, Comment. Math. Helv. **85** (2010) 347–407.

5 Maps, Sheaves and *K*3 Surfaces

5.1 Counting curves

5.1.1 Calabi–Yau 3-folds

A *Calabi–Yau 3-fold* is a non-singular projective variety[1] X of dimension 3 with trivial first Chern class

$$\wedge^3 T_X \cong \mathcal{O}_X.$$

Often the triviality of the fundamental group

$$\pi_1(X) = 1$$

is included in the definition. However, for our purposes, X need not be simply connected.

5.1.2 Maps

Let C be a complete curve with at worst simple nodes as singularities. We do not require C to be connected. The arithmetic genus g of C is defined by the Riemann–Roch formula,

$$\chi(C, \mathcal{O}_C) = 1 - g.$$

We view an algebraic map

$$f : C \to X$$

Lectures on Geometry. Edward Witten, Marc Lackenby, Martin R. Bridson, Helmut Hofer and Rahul Pandharipande.
© Oxford University Press 2017. Published 2017 by Oxford University Press.

that is not constant on *any* connected component of C as parametrizing a subcurve of X. Let

$$\beta = f_*[C] \in H_2(X, \mathbb{Z})$$

be the homology class represented by f. Since f is non-constant, $\beta \neq 0$.

An automorphism of f is an automorphism of the domain

$$\epsilon : C \to C$$

satisfying $f \circ \epsilon = f$. A map f is *stable* [35] if the automorphism group $\mathrm{Aut}(f)$ is finite. Infinite automorphisms can come only from contracted rational and elliptic irreducible components of C incident to too few nodes.

Let $\overline{M}_g(X, \beta)^\bullet$ denote the moduli space of stable maps[2] from genus g curves to X representing the class β. The moduli space $\overline{M}_g(X, \beta)^\bullet$ is a projective Deligne–Mumford stack [5, 17, 35]. Certainly, $\overline{M}_g(X, \beta)^\bullet$ may be singular, non-reduced and disconnected.

The most important structure carried by $\overline{M}_g(X, \beta)^\bullet$ is the obstruction theory [2, 4, 40] governing deformations of maps. The Zariski tangent space at $[f] \in \overline{M}_g(X, \beta)^\bullet$ has dimension

$$\dim_{\mathbb{C}}(T_{[f]}) = 3g - 3 + h^0(C, f^*T_X).$$

The first term on the right corresponds to deformations of the complex structure of C and the second term to deformations of the map with C fixed.[3] The obstruction space is

$$\mathrm{Obs}_{[f]} = H^1(C, f^*T_X).$$

Formally, we may view the moduli space $\overline{M}_g(X, \beta)^\bullet$ as being cut out by $\dim_{\mathbb{C}}(\mathrm{Obs}_{[f]})$ equations in the tangent space. Hence, we expect the dimension of $\overline{M}_g(X, \beta)^\bullet$ to be

$$
\begin{aligned}
\dim_{\mathbb{C}}^{\mathrm{expected}} \left(\overline{M}_g(X, \beta)^\bullet \right) &= 3g - 3 + h^0(C, f^*T_X) - h^1(C, f^*T_X) \\
&= 3g - 3 + \chi(C, f^*T_X) \\
&= 3g - 3 + \int_C c_1(T_X) + \mathrm{rank}_{\mathbb{C}}(T_X)(1 - g) \\
&= 0.
\end{aligned}
$$

The third line is by Riemann–Roch. The Calabi–Yau 3-fold condition is imposed in the fourth line.

Since all curves in Calabi–Yau 3-folds are expected to move in 0-dimensional families, we can hope to count them. While $\overline{M}_g(X, \beta)^\bullet$ may have large positive-dimensional components, the obstruction theory provides a virtual class

$$\left[\overline{M}_g(X, \beta)^\bullet \right]^{\mathrm{vir}} \in H_0(\overline{M}_g(X, \beta), \mathbb{Q})$$

in exactly the expected dimension.

Gromov–Witten theory is the curve counting defined via integration against the virtual class of $\overline{M}_g(X, \beta)^\bullet$. The Gromov–Witten invariants of X are

$$N_{g,\beta}^\bullet = \int_{[\overline{M}_g(X,\beta)^\bullet]^{\mathrm{vir}}} 1 \in \mathbb{Q}.$$

For fixed non-zero $\beta \in H_2(X, \mathbb{Z})$, let

$$Z_{GW,\beta}(u) = \sum_g N_{g,\beta}^\bullet \, u^{2g-2} \in \mathbb{Q}((u))$$

be the partition function.[4] Since $\overline{M}_g(X, \beta)^\bullet$ is empty for g sufficiently negative, $Z_{GW,\beta}(u)$ is a Laurent series.

The Gromov–Witten invariants $N_{g,\beta}^\bullet$ should be viewed as regularized curve counts. The integrals $N_{g,\beta}^\bullet$ are symplectic invariants. A natural idea is to relate the Gromov–Witten invariants to strict symplectic curve counts after perturbing the almost complex structure J. However, analytic difficulties arise. A complete understanding of the symplectic geometry here has not yet been obtained.

5.1.3 Sheaves

We may also approach curve counting in a Calabi–Yau 3-fold X via a gauge/sheaf-theoretic approach [58–60].

We would like to construct a moduli space parametrizing divisors on curves in X. If the subcurve

$$\iota : C \subset X$$

is non-singular, then a divisor determines a line bundle $L \to C$ together with a section $s \in H^0(C, L)$. The associated torsion sheaf

$$\iota_*(L) = F$$

on X has 1-dimensional support and section $s \in H^0(X, F)$. However, for a compact moduli space, we must allow the support curve C to acquire singularities and non-reduced structure. The line bundle L must also be allowed to degenerate.

A *pair* (F, s) consists of a sheaf F on X supported in dimension 1 together with a section $s \in H^0(X, F)$. A pair (F, s) is *stable* if

(i) the sheaf F is pure;

(ii) the section $\mathcal{O}_X \xrightarrow{s} F$ has 0-dimensional cokernel.

Purity here simply means that every non-zero subsheaf of F has support of dimension 1. As a consequence, the scheme-theoretic support $C \subset X$ of F is a Cohen–Macaulay curve.

The support of the cokernel (ii) is a finite-length subscheme $Z \subset C$. If the support C is non-singular, then the stable pair (F, s) is uniquely determined by $Z \subset C$. However, for general C, the subscheme Z does not determine F and s.

The discrete invariants of a stable pair are the holomorphic Euler characteristic $\chi(F) \in \mathbb{Z}$ and the class[5] $[F] \in H_2(X, \mathbb{Z})$. The moduli space $P_n(X, \beta)$ parametrizes stable pairs satisfying

$$\chi(F) = n, \qquad [F] = \beta.$$

After appropriate choices [58], pair stability coincides with stability arising from geo-metric invariant theory [37]. The moduli space $P_n(X, \beta)$ is a therefore a projective scheme.

To define invariants, a virtual cycle is required. The usual deformation theory of pairs is problematic, but the fixed-determinant deformation theory[6] of the associated *complex* in the derived category

$$I^\bullet = \{\mathcal{O}_X \xrightarrow{s} F\} \in D^b(X)$$

is shown in [26, 58] to define a perfect obstruction theory for $P_n(X, \beta)$ of virtual dimen-sion zero. A virtual cycle is then obtained by [2, 4, 40]. The resulting regularized counts are

$$P_{n,\beta} = \int_{[P_n(X,\beta)]^{\mathrm{vir}}} 1 \in \mathbb{Z}.$$

Let

$$Z_{P,\beta}(q) = \sum_{n \in \mathbb{Z}} P_{n,\beta} q^n \in \mathbb{Q}((q))$$

be the partition function. Since $P_n(X, \beta)$ is empty for n sufficient negative, $Z_{P,\beta}(q)$ is a Laurent series.

The Gromov–Witten invariants $N^\bullet_{g,\beta}$ are $\mathbb{Q}$-valued since $\overline{M}_g(X, \beta)^\bullet$ is a Deligne–Mumford stack, but the stable pairs invariants $P_{n,\beta}$ are $\mathbb{Z}$-valued since $P_n(X, \beta)$ is a scheme.

5.2 Correspondence

5.2.1 Two counts

We have seen that there are at least two regularized counting strategies for curves in Calabi–Yau 3-folds. While the Gromov–Witten approach may appear closer to a pure enumerative invariant, since no auxiliary line bundles play a role, in fact the two theories are equivalent!

Conjecture 5.1. *For all Calabi–Yau 3-folds X and non-zero curve classes* $\beta \in H_2(X, \mathbb{Z})$,

$$Z_{GW,\beta}(u) = Z_{P,\beta}(q)$$

after the variable change $-e^{iu} = q$.

Actually, the variable change $-e^{iu} = q$ is not a priori well defined for Laurent series. The issue is addressed by the following rationality property.

Conjecture 5.2. *For all Calabi–Yau 3-folds X and non-zero curve classes* $\beta \in H_2(X, \mathbb{Z})$, *the series* $Z_{P,\beta}(q)$ *is the Laurent expansion of a rational function invariant under* $q \leftrightarrow 1/q$.

In rigid cases, Conjecture 5.1 implies that the contributions of multiple covers in Gromov–Witten theory exactly match the contributions of the divisor choices on thickened curves in the theory of stable pairs. In geometries with moving curves, the meaning of Conjecture 5.1 is more subtle.

5.2.2 Other counts

There are other geometric approaches to curve counting on Calabi–Yau 3-folds. On the map side, a new theory of stability has recently been put forward by B. Kim, A. Kresch and Y.-G. Oh [31], generalizing the well-known theory of admissible covers for dimension-1 targets. On the sheaf side, the older Donaldson–Thomas theory of ideal sheaf counts [13, 63] is very natural to pursue.

While Conjectures 5.1 and 5.2 as stated above are from [58], the relation between Gromov–Witten theory and sheaf counting was first discovered in the context of Donaldson–Thomas theory in [44, 45]. Stable pairs appear to be the closest sheaf enumeration to Gromov–Witten theory. The precise relationship of [31] to the other theories has yet to be discovered, but an equivalence almost surely holds.

5.2.3 Evidence

There are three interesting directions that provide evidence for Conjectures 5.1 and 5.2.

The first is the study of local Calabi–Yau toric surfaces.[7] Both the Gromov–Witten and pairs invariants can be calculated by the virtual localization formula [23]. On the Gromov–Witten side, the topological vertex of [1, 42] evaluates the localization formula. On the stable pairs side, the evaluation is given by box counting [59]. Conjectures 5.1 and 5.2 hold.[8] However, the toric examples are necessarily non-compact. The relevance of toric calculations to compact Calabi–Yau 3-folds is not clear.

The second direction is progress toward a geometric proof of Conjecture 5.2. The obstruction theory for $P_n(X, \beta)$ is self-dual.[9] By results of K. Behrend [3], there exists a constructable function

$$\chi^B : P_n(X, \beta) \to \mathbb{Z}$$

with integral[10] equal to the pairs invariant,

$$\int_{P_n(X,\beta)} \chi^B = P_{n,\beta}.$$

If $P_n(X,\beta)$ is non-singular, then $\chi^B = (-1)^{\dim_{\mathbb{C}}(P_n(X,\beta))}$ is constant and

$$P_{n,\beta} = (-1)^{\dim_{\mathbb{C}}(P_n(X,\beta))} \chi_{\text{top}}(P_n(X,\beta)).$$

In [59], properties of χ^B together with an essential application of Serre duality imply Conjecture 5.2 for irreducible[11] curve classes $\beta \in H_2(X,\beta)$. In remarkable recent work of Y. Toda [64], using variants of Bridgeland's stability conditions, wall-crossing formulas and Serre duality, the rationality of the closely related series

$$Z^\chi_{P,\beta}(q) = \sum_{n \in \mathbb{Z}} \chi_{\text{top}}(P_n(X,\beta))q^n$$

has been proved for *all* non-zero classes $\beta \in H_2(X,\beta)$. A proper inclusion of χ^B into Toda's argument should soon lead to a complete proof of Conjecture 5.2.

The third direction, curve counting on $K3$ surfaces, will be discussed in Sections 5.3 and 5.4. The topic contains a mix of classical and quantum geometry. While there has been recent progress, many beautiful open questions remain.

5.3 Curve counting on *K3* surfaces

5.3.1 Reduced virtual class

Let S be a $K3$ surface and let

$$\beta \in \text{Pic}(S) = H^{1,1}(S,\mathbb{C}) \cap H^2(S,\mathbb{Z})$$

be an non-zero effective curve class.[12] By the virtual dimension formula,

$$\dim_{\mathbb{C}}^{\text{expected}} \left(\overline{M}_g(S,\beta)^\bullet \right) = 3g - 3 + \chi(C, f^*T_S)$$
$$= 3g - 3 + 2(1 - g)$$
$$= g - 1.$$

Let $[f] \in \overline{M}_g(S,\beta)^\bullet$ be a stable map. There is a canonical surjection

$$\text{Obs}_{[f]} \to \mathbb{C} \to 0 \tag{5.1}$$

obtained from the the composition

$$H^1(C, f^*T_S) \cong H^1(C, f^*\Omega_S) \xrightarrow{df} H^1(C, \omega_C) \cong \mathbb{C},$$

where the first isomorphism uses

$$\wedge^2 T_S \cong \mathcal{O}_S.$$

The trivial quotient (5.1) forces the vanishing of $\left[\overline{M}_g(S,\beta)^\bullet \right]^{\text{vir}}$. However, the obstruction theory can be modified to reduce the obstruction space to the kernel of (5.1). A reduced virtual class

$$\left[\overline{M}_g(S,\beta)^\bullet \right]^{\text{red}} \in H_{2g}\left(\overline{M}_g(S,\beta)^\bullet, \mathbb{Q} \right), \tag{5.2}$$

in dimension 1 greater than expected, is therefore defined.

By constructing trivial quotients of $\text{Obs}_{[f]}$ for each connected component of the domain, the reduced virtual class (5.2) is easily seen to be supported on the locus of curves with connected domains. Hence, we need only consider

$$\overline{M}_g(S,\beta) \subset \overline{M}_g(S,\beta)^\bullet.$$

We can also consider stable maps from r-pointed curves. The pointed moduli space $\overline{M}_{g,r}(S,\beta)$ has a reduced virtual class of dimension $g + r$.

5.3.2 Descendents

The reduced Gromov–Witten theory of S is defined via integration against $\left[\overline{M}_{g,r}(S,\beta) \right]^{\text{red}}$. Let

$$\text{ev}_i : \overline{M}_{g,r}(S,\beta) \to S,$$

$$L_i \to \overline{M}_{g,r}(S,\beta)$$

denote the evaluation maps and cotangent lines bundles associated to the r marked points. Let $\gamma_1, \ldots, \gamma_m$ be a basis of $H^*(S,\mathbb{Q})$ and let

$$\psi_i = c_1(L_i) \in \overline{M}_{g,r}(S,\beta).$$

The *descendent* fields, denoted in the brackets by $\tau_k(\gamma_j)$, correspond to the classes $\psi_i^k \cup \text{ev}_i^*(\gamma_j)$ on the moduli space of maps. Let

$$\left\langle \tau_{k_1}(\gamma_{l_1}) \cdots \tau_{k_r}(\gamma_{l_r}) \right\rangle_{g,\beta}^{\text{red}} = \int_{[\overline{M}_{g,r}(S,\beta)]^{\text{red}}} \prod_{i=1}^{r} \psi_i^{k_i} \cup \text{ev}_i^*(\gamma_{l_i}) \tag{5.3}$$

denote the descendent Gromov–Witten invariants. Of course, (5.3) vanishes if the integrand does not match the dimension of the reduced virtual class.

The reduced Gromov–Witten theory is invariant under deformations of S that preserve β as an algebraic class. A standard argument[13] shows that the invariant (5.3) depends *only* on the norm

$$\langle \beta, \beta \rangle = \int_S \beta \cup \beta$$

and the divisibility of $\beta \in H^2(S, \mathbb{Z})$.

Let us now specialize, for the remainder of Section 5.3.2, to an elliptically fibered $K3$ surface

$$\nu : S \to \mathbb{P}^1$$

with a section. We assume the section and fiber classes

$$\mathbf{s}, \mathbf{f} \in H^2(S, \mathbb{Z})$$

span $\mathrm{Pic}(S)$. The cone of effective curve classes is

$$V = \{m\mathbf{s} + n\mathbf{f} \mid m \geq 0,\ n \geq 0,\ (m, n) \neq (0, 0)\}.$$

Since the norm of $d\mathbf{s} + dk\mathbf{f}$ is $2d^2(k-1)$, effective classes with all divisibilities $d \geq 1$ and norms at least $-2d^2$ can be found on S. Elementary arguments show the integrals (5.3) vanish in all other cases.[14]

A natural descendent potential function for the reduced theory of $K3$ surfaces is defined by

$$F_{g,m}^S\big(\tau_{k_1}(\gamma_{l_1}) \cdots \tau_{k_r}(\gamma_{l_r})\big) = \sum_{n=0}^{\infty} \big\langle \tau_{k_1}(\gamma_{l_1}) \cdots \tau_{k_r}(\gamma_{l_r}) \big\rangle_{g, m\mathbf{s}+n\mathbf{f}}^{\mathrm{red}} q^{m(n-m)}$$

for $g \geq 0$ and $m \geq 1$. The following conjecture is made jointly with D. Maulik.

Conjecture 5.3. $F_{g,m}^S\big(\tau_{k_1}(\gamma_{l_1}) \cdots \tau_{k_r}(\gamma_{l_r})\big)$ *is the Fourier expansion in q of a quasimodular form of level m^2 with pole at $q = 0$ of order at most m^2.*

By the ring of quasimodular forms of level m^2 with possible poles at $q = 0$, we mean the algebra generated by the Eisenstein series[15] E_2 over the ring of modular forms of level m^2. In [50], Conjecture 5.3 is proved in the primitive case $m = 1$ by relations in the moduli of curves [15], degeneration methods [47] and the elliptic curve results of [51–53]. The $m > 1$ case appears to require new techniques.

Let $[p] \in H^4(S, \mathbb{Z})$ denote the Poincaré dual of a point. The simplest of the $K3$ series is the count of genus g curves passing through g points,

$$F_{g,1}^S\big(\tau_0(p) \cdots \tau_0(p)\big) = \eta^{-24} \left(-\frac{1}{24} q \frac{d}{dq} E_2 \right)^g,$$

calculated[16] by J. Bryan and C. Leung [10]. Here

$$\eta(q) = q^{1/24} \prod_{n=1}^{\infty} (1 - q^n)$$

is Dedekind's function. Similar calculations in genus 1 for $m = 2$ have been done in [39].

5.3.3 λ_g integrals for $K3$ surfaces

A connection to the enumerative geometry of Calabi–Yau 3-folds holds for special integrals in the reduced Gromov–Witten theory of $K3$ surfaces. Let

$$R_{g,\beta} = \int_{[\overline{M}_g(S,\beta)]^{red}} (-1)^g \lambda_g \tag{5.4}$$

for effective curve classes $\beta \in H^2(S, \mathbb{Z})$. Here, the integrand λ_g is the top Chern class of the Hodge bundle

$$\mathbb{E}_g \to \overline{M}_g(S, \beta)$$

with fiber $H^0(C, \omega_C)$ over moduli point

$$[f : C \to S] \in \overline{M}_g(S, \beta).$$

See [14, 23] for a discussion of Hodge classes in Gromov–Witten theory.

The integrals (5.4) arise from the following 3-fold geometry. Let

$$\pi : X \to \mathbb{P}^1$$

be a $K3$-fibered Calabi–Yau 3-fold with

$$\iota : S \xrightarrow{\sim} \pi^{-1}(0) \subset X.$$

Assume further that the family of $K3$ surfaces determined by X is transverse to the Noether–Lefschetz divisor in the moduli of $K3$ surfaces along which β is an algebraic class. Then, the moduli space

$$\overline{M}_g(S, \beta) \subset \overline{M}_g(X, \iota_*\beta)$$

is a connected component. The integral (5.4) is precisely the contribution of $\overline{M}_g(S, \beta)$ to the Gromov–Witten theory of X [49]. The discussion here may be viewed as an algebraic analogue of the twistor construction of [10].

The definition of the BPS counts[17] associated to the Hodge integrals (5.4) is straightforward. Let $\alpha \in \text{Pic}(S)$ be a effective primitive class The Gromov–Witten potential $F_\alpha(u, v)$ for classes proportional to α is

$$F_\alpha = \sum_{g=0}^{\infty} \sum_{m=0}^{\infty} R_{g,m\alpha} u^{2g-2} v^{m\alpha}.$$

The BPS counts $r_{g,m\alpha}$ are uniquely defined by the following equation:

$$F_\alpha = \sum_{g=0}^{\infty} \sum_{m=0}^{\infty} r_{g,m\alpha} u^{2g-2} \sum_{d>0} \frac{1}{d} \left(\frac{\sin(du/2)}{u/2} \right)^{2g-2} v^{dm\alpha}.$$

We have defined BPS counts for both primitive and divisible classes.

The string-theoretic calculations of S. Katz, A. Klemm and C. Vafa [28] via heterotic duality yield two conjectures.

Conjecture 5.4. *The BPS count $r_{g,\beta}$ depends upon β only through the norm $\langle \beta, \beta \rangle$.*

Assuming the validity of Conjecture 5.4, let $r_{g,h}$ denote the BPS count associated to a class β satisfying

$$\langle \beta, \beta \rangle = 2h - 2.$$

Conjecture 5.4 is rather surprising from the point of view of Gromov–Witten theory. The invariants $R_{g,\beta}$ depend a priori upon both the norm and the divisibility of β.

Conjecture 5.5. *The BPS counts $r_{g,h}$ are uniquely determined by the following equation:*

$$\sum_{g=0}^{\infty} \sum_{h=0}^{\infty} (-1)^g r_{g,h} \left(\sqrt{z} - \frac{1}{\sqrt{z}} \right)^{2g} q^h = \prod_{n=1}^{\infty} \frac{1}{(1 - q^n)^{20}(1 - zq^n)^2(1 - z^{-1}q^n)^2}.$$

As a consequence of Conjecture 5.5, $r_{g,h}$ vanishes if $g > h$ and

$$r_{g,g} = (-1)^g (g + 1).$$

The first values are as follows:

$r_{g,h}$	$h = 0$	1	2	3	4
$g = 0$	1	24	324	3200	25 650
1		−2	−54	−800	−8550
2			3	88	1401
3				−4	−126
4					5

Conjectures 5.4 and 5.5 provide a complete solution for λ_g integrals in the reduced Gromov–Witten theory of $K3$ surfaces. The answer is compatible with Conjecture 5.3, as expected, since Hodge integrals may be expressed in terms of descendent integrals [14].

5.3.4 Stable pairs on *K*3 surfaces

Let S be a $K3$ surface with an irreducible class $\beta \in H^2(S, \mathbb{Z})$ satisfying

$$\langle \beta, \beta \rangle = 2h - 2,$$

and let $P_n(S, h)$ denote the associated moduli space of pairs on S. Consider again the $K3$-fibered Calabi–Yau 3-fold

$$\pi : X \to \mathbb{P}^1.$$

A deformation argument in [60] proves that

$$P_n(S, h) \subset P_n(X, \iota_* \beta) \tag{5.5}$$

is a connected component of the moduli space of stable pairs of X. Moreover, $P_n(S, h)$ is a *non-singular* projective variety [30, 60] of dimension $n + 2h - 1$.

Let Ω_P be the cotangent bundle of the moduli space $P_n(S, h)$. The self-dual obstruction theory on $P_n(S, h)$ induced from the inclusion (5.5) has obstruction bundle Ω_P. Hence, the contribution of $P_n(S, h)$ to the stable pairs invariants of X is

$$Z^S_{P,h}(y) = \sum_n \int_{P_n(S,h)}^{\!\!\!\!\!} c_{n+2h-1}(\Omega_P) y^n$$

$$= \sum_n (-1)^{n+2h-1} e(P_n(S, h)) y^n.$$

Here, we have written the stable pairs partition function in the variable y instead of the traditional q, since the latter will be reserved for the Fourier expansions of modular forms.[18]

Fortunately, the topological Euler characteristics of $P_n(S, h)$ have been calculated by T. Kawai and K. Yoshioka. By Theorem 5.80 of [30],

$$\sum_{h=0}^{\infty} \sum_{n=1-h}^{\infty} e(P_n(S, h)) y^n q^h =$$

$$\left(\sqrt{y} - \frac{1}{\sqrt{y}} \right)^{-2} \prod_{n=1}^{\infty} \frac{1}{(1 - q^n)^{20}(1 - yq^n)^2(1 - y^{-1}q^n)^2}.$$

For our pairs invariants, we require the signed Euler characteristics,

$$\sum_{h=0}^{\infty} Z^S_h(y) q^h = \sum_{h=0}^{\infty} \sum_{n=1-h}^{\infty} (-1)^{n+2h-1} e(P_n(S, h)) y^n q^h.$$

Therefore, $\sum_{h=0}^{\infty} Z^S_{P,h}(y) q^h$ equals

$$-\left(\sqrt{-y} - \frac{1}{\sqrt{-y}} \right)^{-2} \prod_{n=1}^{\infty} \frac{1}{(1 - q^n)^{20}(1 + yq^n)^2(1 + y^{-1}q^n)^2}.$$

5.3.5 Correspondence

We are now in a position to check whether the Katz–Klemm–Vafa predictions for the λ_g integrals in the reduced Gromov–Witten theory of S are compatible with the above stable pairs calculations via the maps/pairs correspondence of Conjecture 5.1.

In the β-irreducible case, the Gromov–Witten partition function takes the form

$$\sum_{h=0}^{\infty} Z_{GW,h}^S(u)\, q^h = \sum_{g=0}^{\infty} \sum_{h=0}^{\infty} r_{g,h} u^{2g-2} \left(\frac{\sin(u/2)}{u/2} \right)^{2g-2} q^h.$$

After substituting $-e^{iu} = y$, we find

$$\sum_{h=0}^{\infty} Z_{GW,h}^S(y)\, q^h = \sum_{g=0}^{\infty} \sum_{h=0}^{\infty} (-1)^{g-1} r_{g,h} \left(\sqrt{-y} - \frac{1}{\sqrt{-y}} \right)^{2g-2} q^h.$$

By Conjecture 5.5, $\sum_{h=0}^{\infty} Z_{GW,h}^S(y)\, q^h$ equals

$$-\left(\sqrt{-y} - \frac{1}{\sqrt{-y}} \right)^{-2} \prod_{n=1}^{\infty} \frac{1}{(1-q^n)^{20}(1+yq^n)^2(1+y^{-1}q^n)^2}$$

which is $\sum_{h=0}^{\infty} Z_{P,h}^S(y)\, q^h$.

The maps/pairs correspondence of Conjecture 5.1 therefore works perfectly, assuming the Katz–Klemm–Vafa prediction for the reduced Gromov–Witten theory. But can the Katz–Klemm–Vafa prediction for stable maps be proved? The answer is yes in genus 0. The proof is our last topic.

5.4 The Yau–Zaslow conjecture

5.4.1 Genus 0

The genus 0 parts of Conjectures 5.4 and 5.5 for $K3$ surfaces were predicted earlier by S.-T. Yau and E. Zaslow [66].

Conjecture 5.6. *The BPS count $r_{0,\beta}$ depends upon β only through the norm $\langle \beta, \beta \rangle$.*

Let $r_{0,m,h}$ denote the genus 0 BPS count associated to a class β of divisibility m satisfying

$$\langle \beta, \beta \rangle = 2h - 2.$$

Assuming that Conjecture 5.6 holds, we define

$$r_{0,h} = r_{0,m,h}$$

independent[19] of m.

Conjecture 5.7. *The BPS counts $r_{0,h}$ are uniquely determined by*

$$\sum_{h\geq 0} r_{0,h}\, q^{h-1} = q^{-1}\prod_{n=1}^{\infty}(1-q^n)^{-24}. \tag{5.6}$$

A mathematical derivation of the Yau–Zaslow conjectures for primitive classes β via Euler characteristics of compactified Jacobians following [66] can be found in [6, 11, 16]. The Yau–Zaslow formula (5.6) was proved via Gromov–Witten theory for primitive classes β by J. Bryan and C. Leung [10]. An early calculation by A. Gathmann [20] for a class β of divisibility 2 was important for the correct formulation of the conjectures. Conjectures 5.6 and 5.7 have been proved in the divisibility-2 case by J. Lee and C. Leung [38] and B. Wu [65].

The main result of the paper [34] with A. Klemm, D. Maulik and E. Scheidegger is a proof of Conjectures 5.6 and 5.7 in all cases.

Theorem 5.8. *The Yau–Zaslow conjectures hold for all non-zero effective classes $\beta \in \mathrm{Pic}(S)$ on a K3 surface S.*

The proof, using the connection to Noether–Lefschetz theory [49], mirror symmetry, and modular form identities, is surveyed in Sections 5.4.2–5.4.5.

5.4.2 Noether–Lefschetz theory

5.4.2.1 K3 lattice

Let S be a K3 surface. The second cohomology of S is a rank 22 lattice with intersection form

$$H^2(S,\mathbb{Z}) \cong U \oplus U \oplus U \oplus E_8(-1) \oplus E_8(-1), \tag{5.7}$$

where

$$U = \begin{pmatrix} 0 & 1 \\ 1 & 0 \end{pmatrix}$$

and

$$E_8(-1) = \begin{pmatrix} -2 & 0 & 1 & 0 & 0 & 0 & 0 & 0 \\ 0 & -2 & 0 & 1 & 0 & 0 & 0 & 0 \\ 1 & 0 & -2 & 1 & 0 & 0 & 0 & 0 \\ 0 & 1 & 1 & -2 & 1 & 0 & 0 & 0 \\ 0 & 0 & 0 & 1 & -2 & 1 & 0 & 0 \\ 0 & 0 & 0 & 0 & 1 & -2 & 1 & 0 \\ 0 & 0 & 0 & 0 & 0 & 1 & -2 & 1 \\ 0 & 0 & 0 & 0 & 0 & 0 & 1 & -2 \end{pmatrix}$$

is the (negative) Cartan matrix. The intersection form (5.7) is even.

5.4.2.2 Lattice polarization

A primitive[20] class $L \in \text{Pic}(S)$ is a *quasipolarization* if

$$\langle L, L \rangle > 0 \quad \text{and} \quad \langle L, [C] \rangle \geq 0$$

for every curve $C \subset S$. A sufficiently high tensor power L^n of a quasipolarization is base point free and determines a birational morphism

$$S \to \tilde{S}$$

contracting A-D-E configurations of (-2)-curves on S. Hence, every quasipolarized $K3$ surface is algebraic.

Let Λ be a fixed-rank r primitive[21] embedding

$$\Lambda \subset U \oplus U \oplus U \oplus E_8(-1) \oplus E_8(-1)$$

with signature $(1, r-1)$, and let $v_1, \ldots, v_r \in \Lambda$ be an integral basis. The discriminant is

$$\Delta(\Lambda) = (-1)^{r-1} \det \begin{pmatrix} \langle v_1, v_1 \rangle & \cdots & \langle v_1, v_r \rangle \\ \vdots & \ddots & \vdots \\ \langle v_r, v_1 \rangle & \cdots & \langle v_r, v_r \rangle \end{pmatrix}.$$

The sign is chosen so $\Delta(\Lambda) > 0$.

A Λ-*polarization* of a $K3$ surface S is a primitive embedding

$$j : \Lambda \to \text{Pic}(S)$$

satisfying two properties:

(i) The lattice pairs $\Lambda \subset U^3 \oplus E_8(-1)^2$ and $\Lambda \subset H^2(S, \mathbb{Z})$ are isomorphic via an isometry that restricts to the identity on Λ.

(ii) $\text{Im}(j)$ contains a quasipolarization.

By (ii), every Λ-polarized $K3$ surface is algebraic.

The period domain M of Hodge structures of type $(1, 20, 1)$ on the lattice $U^3 \oplus E_8(-1)^2$ is an analytic open set of the 20-dimensional non-singular isotropic quadric Q:

$$M \subset Q \subset \mathbb{P}\big((U^3 \oplus E_8(-1)^2) \otimes_{\mathbb{Z}} \mathbb{C}\big).$$

Let $M_\Lambda \subset M$ be the locus of vectors orthogonal to the entire sublattice $\Lambda \subset U^3 \oplus E_8(-1)^2$.

Let Γ be the isometry group of the lattice $U^3 \oplus E_8(-1)^2$, and let

$$\Gamma_\Lambda \subset \Gamma$$

be the subgroup restricting to the identity on Λ. By global Torelli, the moduli space $\mathcal{M}_\Lambda$ of Λ-polarized *K3* surfaces is the quotient

$$\mathcal{M}_\Lambda = M_\Lambda / \Gamma_\Lambda.$$

We refer the reader to [12] for a detailed discussion.

5.4.2.3 Families

Let X be a compact 3-dimensional complex manifold equipped with holomorphic line bundles

$$L_1, \ldots, L_r \to X$$

and a holomorphic map

$$\pi : X \to C$$

to a non-singular complete curve.

The tuple $(X, L_1, \ldots, L_r, \pi)$ is a *1-parameter family of non-singular Λ-polarized K3 surfaces* if

(i) the fibers $(X_\xi, L_{1,\xi}, \ldots, L_{r,\xi})$ are Λ-polarized *K3* surfaces via

$$v_i \mapsto L_{i,\xi}$$

for every $\xi \in C$;

(ii) there exists a $\lambda^\pi \in \Lambda$ that is a quasipolarization of all fibers of π simultaneously.

The family π yields a morphism

$$\iota_\pi : C \to \mathcal{M}_\Lambda$$

to the moduli space of Λ-polarized *K3* surfaces.

Let $\lambda^\pi = \lambda_1^\pi v_1 + \ldots + \lambda_r^\pi v_r$. A vector $(d_1, \ldots, d_r)$ of integers is *positive* if

$$\sum_{i=1}^r \lambda_i^\pi d_i > 0.$$

If $\beta \in \mathrm{Pic}(X_\xi)$ has intersection numbers

$$d_i = \langle L_{i,\xi}, \beta \rangle,$$

then β has positive degree with respect to the quasipolarization if and only if $(d_1, \ldots, d_r)$ is positive.

5.4.2.4 Noether–Lefschetz divisors

Noether–Lefschetz numbers are defined in [49] by the intersection of $\iota_\pi(C)$ with Noether–Lefschetz divisors in $\mathcal{M}_\Lambda$. We briefly review the definition of the Noether–Lefschetz divisors.

Let $(\mathbb{L}, \iota)$ be a rank $r + 1$ lattice $\mathbb{L}$ with an even symmetric bilinear form $\langle \cdot, \cdot \rangle$ and a primitive embedding

$$\iota : \Lambda \to \mathbb{L}.$$

Two data sets $(\mathbb{L}, \iota)$ and $(\mathbb{L}', \iota')$ are isomorphic if there is an isometry that restricts to the identity on Λ. The first invariant of the data $(\mathbb{L}, \iota)$ is the discriminant $\Delta \in \mathbb{Z}$ of $\mathbb{L}$.

An additional invariant of $(\mathbb{L}, \iota)$ can be obtained by considering any vector $v \in \mathbb{L}$ for which

$$\mathbb{L} = \iota(\Lambda) \oplus \mathbb{Z}v. \tag{5.8}$$

The pairing

$$\langle v, \cdot \rangle : \Lambda \to \mathbb{Z}$$

determines an element of $\delta_v \in \Lambda^*$. Let $G = \Lambda^*/\Lambda$ be the quotient defined via the injection $\Lambda \to \Lambda^*$ obtained from the pairing $\langle \cdot, \cdot \rangle$ on Λ. The group G is abelian of order equal to the discriminant $\Delta(\Lambda)$. The image

$$\delta \in G/\pm$$

of δ_v is easily seen to be independent of v satisfying (5.8). The invariant δ is the *coset* of $(\mathbb{L}, \iota)$

By elementary arguments, two data sets $(\mathbb{L}, \iota)$ and $(\mathbb{L}', \iota')$ of rank $r + 1$ are isomorphic if and only if the discriminants and cosets are equal.

Let $v_1, \ldots, v_r$ be an integral basis of Λ as before. The pairing of $\mathbb{L}$ with respect to an extended basis $v_1, \ldots, v_r, v$ is encoded in the matrix

$$\mathbb{L}_{h,d_1,\ldots,d_r} = \begin{pmatrix} \langle v_1, v_1 \rangle & \cdots & \langle v_1, v_r \rangle & d_1 \\ \vdots & \ddots & \vdots & \vdots \\ \langle v_r, v_1 \rangle & \cdots & \langle v_r, v_r \rangle & d_r \\ d_1 & \cdots & d_r & 2h - 2 \end{pmatrix}.$$

The discriminant is

$$\Delta(h, d_1, \ldots, d_r) = (-1)^r \det(\mathbb{L}_{h,d_1,\ldots,d_r}).$$

The coset $\delta(h, d_1, \ldots, d_r)$ is represented by the functional

$$v_i \mapsto d_i.$$

The Noether–Lefschetz divisor $P_{\Delta, \delta} \subset \mathcal{M}_\Lambda$ is the closure of the locus of Λ-polarized K3 surfaces S for which $(\mathrm{Pic}(S), j)$ has rank $r + 1$, discriminant Δ and coset δ. By the Hodge index theorem, $P_{\Delta, \delta}$ is empty unless $\Delta > 0$.

Let $h, d_1, \ldots, d_r$ determine a positive discriminant

$$\Delta(h, d_1, \ldots, d_r) > 0.$$

The Noether–Lefschetz divisor $D_{h, (d_1, \ldots, d_r)} \subset \mathcal{M}_\Lambda$ is defined by the weighted sum

$$D_{h, (d_1, \ldots, d_r)} = \sum_{\Delta, \delta} m(h, d_1, \ldots, d_r | \Delta, \delta) \cdot [P_{\Delta, \delta}]$$

where the multiplicity $m(h, d_1, \ldots, d_r | \Delta, \delta)$ is the number of elements β of the lattice $(\mathbb{L}, \iota)$ of type (Δ, δ) satisfying

$$\langle \beta, \beta \rangle = 2h - 2, \qquad \langle \beta, v_i \rangle = d_i. \tag{5.9}$$

If the multiplicity is non-zero, then $\Delta | \Delta(h, d_1, \ldots, d_r)$, so only finitely many divisors appear in the above sum.

If $\Delta(h, d_1, \ldots, d_r) = 0$, the divisor $D_{h, (d_1, \ldots, d_r)}$ has an alternate definition. The tautological line bundle $\mathcal{O}(-1)$ is Γ-equivariant on the period domain M_Λ and descends to the *Hodge line bundle*

$$\mathcal{K} \to \mathcal{M}_\Lambda.$$

We define $D_{h, (d_1, \ldots, d_r)} = \mathcal{K}^*$. See [49] for an alternate view of degenerate intersection.

If $\Delta(h, d_1, \ldots, d_r) < 0$ then the divisor $D_{h, (d_1, \ldots, d_r)}$ on $\mathcal{M}_\Lambda$ is defined to vanish by the Hodge index theorem.

5.4.2.5 Noether–Lefschetz numbers

Let Λ be a lattice of discriminant $l = \Delta(\Lambda)$ and let $(X, L_1, \ldots, L_r, \pi)$ be a 1-parameter family of Λ-polarized K3 surfaces. The Noether–Lefschetz number $NL^\pi_{h, d_1, \ldots, d_r}$ is the classical intersection product

$$NL^\pi_{h, (d_1, \ldots, d_r)} = \int_C \iota^*_\pi [D_{h, (d_1, \ldots, d_r)}]. \tag{5.10}$$

Let $\mathrm{Mp}_2(\mathbb{Z})$ be the metaplectic double cover of $SL_2(\mathbb{Z})$. There is a canonical representation [7] associated to Λ,

$$\rho^*_\Lambda : \mathrm{Mp}_2(\mathbb{Z}) \to \mathrm{End}(\mathbb{C}[G]).$$

The full set of Noether–Lefschetz numbers $NL^\pi_{h,d_1,\dots,d_r}$ defines a vector-valued modular form

$$\Phi^\pi(q) = \sum_{\gamma \in G} \Phi^\pi_\gamma(q) v_\gamma \in \mathbb{C}[[q^{\frac{1}{2}}]] \otimes \mathbb{C}[G],$$

of weight $(22 - r)/2$ and type ρ^*_Λ by results[22] of Borcherds and Kudla–Millson [7, 36]. The Noether–Lefschetz numbers are the coefficients[23] of the components of Φ^π,

$$NL^\pi_{h,(d_1,\dots,d_r)} = \Phi^\pi_\gamma \left[\frac{\Delta(h, d_1, \dots, d_r)}{2l} \right],$$

where $\delta(h, d_1, \dots, d_r) = \pm\gamma$. The modular form results significantly constrain the Noether–Lefschetz numbers.

5.4.2.6 Refinements

If $d_1, \dots, d_r$ do not simultaneously vanish, refined Noether–Lefschetz divisors are defined. If $\Delta(h, d_1, \dots, d_r) > 0$, then

$$D_{m,h,(d_1,\dots,d_r)} \subset D_{h,(d_1,\dots,d_r)}$$

is defined by requiring the class $\beta \in \mathrm{Pic}(S)$ to satisfy (5.9) and have divisibility $m > 0$. If $\Delta(h, d_1, \dots, d_r) = 0$, then

$$D_{m,h,(d_1,\dots,d_r)} = D_{h,(d_1,\dots,d_r)}$$

if $m > 0$ is the greatest common divisor of $d_1, \dots, d_r$ and 0 otherwise.

Refined Noether–Lefschetz numbers are defined by

$$NL^\pi_{m,h,(d_1,\dots,d_r)} = \int_C \iota^*_\pi [D_{m,h,(d_1,\dots,d_r)}]. \tag{5.11}$$

The full set of Noether–Lefschetz numbers $NL^\pi_{h,(d_1,\dots,d_r)}$ is easily shown in [34] to determine the refined numbers $NL^\pi_{m,h,(d_1,\dots,d_r)}$.

5.4.3 Three theories

The main geometric idea in the proof of Theorem 5.8 is the relationship of three theories associated to a 1-parameter family

$$\pi : X \to C$$

of Λ-polarized $K3$ surfaces:

(i)   the Noether–Lefschetz numbers of π;

(ii) the genus 0 Gromov–Witten invariants of X,

(iii) the genus 0 reduced Gromov–Witten invariants of the $K3$ fibers.

The Noether–Lefschetz numbers (i) are classical intersection products, while the Gromov–Witten invariants (ii) and (iii) are quantum in origin. For (ii), we view the theory in terms the Gopakumar–Vafa invariants [21, 22].

Let $n^X_{0,(d_1,\ldots,d_r)}$ denote the Gopakumar–Vafa invariant of X in genus 0 for π-vertical curve classes of degrees $d_1,\ldots,d_r$ with respect to the line bundles $L_1,\ldots,L_r$. Let $r_{0,m,h}$ denote the reduced $K3$ invariant. The following result is proved[24] in [49] by a comparison of the reduced and usual deformation theories of maps of curves to the $K3$ fibers of π.

Theorem 5.9. *For degrees $(d_1,\ldots,d_r)$ positive with respect to the quasi-polarization λ^π,*

$$n^X_{0,(d_1,\ldots,d_r)} = \sum_{h=0}^{\infty}\sum_{m=1}^{\infty} r_{0,m,h} \cdot NL^\pi_{m,h,(d_1,\ldots,d_r)}.$$

5.4.4 The STU model

The STU model[25] is a particular non-singular projective Calabi–Yau 3-fold X equipped with a fibration

$$\pi : X \to \mathbb{P}^1. \tag{5.12}$$

Except for 528 points $\xi \in \mathbb{P}^1$, the fibers

$$X_\xi = \pi^{-1}(\xi)$$

are non-singular elliptically fibered $K3$ surfaces. The 528 singular fibers X_ξ have exactly 1 ordinary double point singularity each.

The 3-fold X is constructed as a non-singular anticanonical section of the non-singular projective toric 4-fold Y defined by 10 rays with primitives

$$\rho_1 = (1,0,2,3), \qquad \rho_2 = (-1,0,2,3),$$
$$\rho_3 = (0,1,2,3), \qquad \rho_4 = (0,-1,2,3),$$
$$\rho_5 = (0,0,2,3), \qquad \rho_6 = (0,0,-1,0), \qquad \rho_7 = (0,0,0,-1),$$
$$\rho_8 = (0,0,1,2), \qquad \rho_9 = (0,0,0,1), \qquad \rho_{10} = (0,0,1,1).$$

The Picard rank of Y is 6. The fibration (5.12) is obtained from a non-singular toric fibration

$$\pi^Y : Y \to \mathbb{P}^1.$$

The image of

$$\mathrm{Pic}(Y) \to \mathrm{Pic}(X_\xi)$$

determines a rank 2 sublattice of each fiber $\mathrm{Pic}(X_\xi)$ with intersection form

$$\Lambda = \begin{pmatrix} 0 & 1 \\ 1 & 0 \end{pmatrix}.$$

Let $L_1, L_2 \to X$ denote line bundles that span the standard basis of the form Λ after restriction.

Strictly speaking, the tuple (X, L_1, L_2, π) is not a 1-parameter family of Λ-polarized K3 surfaces. The only failing is the 528 singular fibers of π. Let

$$\epsilon : C \xrightarrow{2\text{-}1} \mathbb{P}^1$$

be a hyperelliptic curve branched over the 528 points of $\mathbb{P}^1$ corresponding to the singular fibers of π. The family

$$\epsilon^*(X) \to C$$

has 3-fold double point singularities over the 528 nodes of the fibers of the original family. Let

$$\tilde{\pi} : \tilde{X} \to C$$

be obtained from a small resolution

$$\tilde{X} \to \epsilon^*(X).$$

Let $\tilde{L}_i \to \tilde{X}$ be the pullback of L_i by ϵ. The data

$$(\tilde{X}, \tilde{L}_1, \tilde{L}_2, \tilde{\pi})$$

determine a 1-parameter family of Λ-polarized K3 surfaces; see Section 5.3 of [49]. The simultaneous quasipolarization is obtained from the projectivity of X.

5.4.5 Proof of Theorem 5.8

Theorem 5.8 is proved in [34] by studying Theorem 5.9 applied to the STU model. There are four basic steps:

(i) The modular form [7, 36] determining the intersections of the base curve with the Noether–Lefschetz divisors is calculated. For the STU model, the modular form has vector dimension 1 and is proportional to the product $E_4 E_6$ of Eisenstein series.

(ii) Theorem 5.9 is used to show that the 3-fold BPS counts $n^{\widetilde{X}}_{0,(d_1,d_2)}$ then *determine* all the reduced *K*3 invariants $r_{0,m,h}$. Strong use is made of the rank 2 lattice of the STU model.

(iii) The BPS counts $n^{\widetilde{X}}_{0,(d_1,d_2)}$ are calculated via mirror symmetry. Since the STU model is realized as a Calabi–Yau complete intersection in a non-singular toric variety, the genus 0 Gromov–Witten invariants are obtained after proved mirror transformations from hypergeometric series [18, 19, 41]. The Klemm–Lerche–Mayr identity, proved in [34], shows that the invariants $n^{\widetilde{X}}_{0,(d_1,d_2)}$ are themselves related to modular forms.

(iv) Theorem 5.8 then follows from the Harvey–Moore identity, which simultaneously relates the modular structures of

$$n^{\widetilde{X}}_{0,(d_1,d_2)}, \quad r_{0,m,h} \quad \text{and} \quad NL^{\widetilde{\pi}}_{m,h,(d_1,d_2)}$$

in the form specified by Theorem 5.9.

The Harvey–Moore identity of part (iv) is simple to state. Let

$$f(\tau) = \frac{E_4(\tau)E_6(\tau)}{\eta(\tau)^{24}} = \sum_{n=-1}^{\infty} c(n)q^n,$$

where $q = \exp(2\pi i \tau)$. Then,

$$\frac{f(\tau_1)E_4(\tau_2)}{j(\tau_1) - j(\tau_2)} = \frac{q_1}{q_1 - q_2} + E_4(\tau_2) - \sum_{d,k,\ell>0} \ell^3 c(k\ell) q_1^{kd} q_2^{\ell d}. \tag{5.13}$$

Equation (5.13) was conjectured in [24] and proved by D. Zagier—the proof is presented in Section 4 of [34].

The strategy of the proof of the Yau–Zaslow conjectures is special to genus 0. Much less is known for higher genus. For genus 1, the Katz–Klemm–Vafa conjectures follow for all classes on *K*3 surfaces from the Yau–Zaslow conjectures via the boundary relation for λ_1 in the moduli of elliptic curves. For genus 2 and 3, A. Pixton [62] has proved the Katz–Klemm–Vafa formula for primitive classes using boundary relations for λ_2 and λ_3 on $\overline{M}_2$ and $\overline{M}_3$, respectively. New ideas will be required for a complete proof of Conjectures 5.4 and 5.5.

Acknowledgments

The paper accompanied my lecture at the Clay Research Conference in Cambridge, Massachusetts in May 2008. The discussion of the enumerative geometry of stable pairs in Sections 5.1 and 5.2 reflects joint work with R. Thomas. The study of descendent integrals in the reduced Gromov–Witten theory of *K*3 surfaces in Section 5.3 is joint work with D. Maulik. The proof of the Yau–Zaslow conjectures reported in Section 5.4 is joint work with A. Klemm, D. Maulik, and E. Scheidegger. My research was partially supported by NSF Grant DMS-0500187.

Note

Many of the ideas discussed here are valid in more general contexts. For example, the stable maps/pairs correspondence is conjectured in [58] for all 3-folds—the Calabi–Yau condition is not necessary. The $K3$ study can be pursued along similar lines for abelian surfaces; see [9] for a start. The Enriques surface is a close cousin [48].

Updates in 2016

As this chapter is appearing some time after the lecture, some comments on progress in the field are in order. Concerning the state of affairs of the Gromov–Witten/pairs conjectures:

- Conjecture 5.1 has been proved for a wide class of Calabi–Yau 3-folds (including the quintic 3-fold) that admit good degenerations [57]. The extension of the Gromov–Witten/pairs correspondences to descendent invariants plays a central role [54, 56].
- Conjecture 5.2 has been proved for all Calabi–Yau 3-folds by Toda [64] and Bridgeland [8], as indicated in Section 5.2.3. However, for more general 3-folds, the rationality of the stable pairs descendent series is still open; see [55] for toric results.

Concerning the Katz–Klemm–Vafa conjectures:

- In the primitive case, a proof is given in [50] using a reduced Gromov–Witten/pairs correspondence for the non-compact 3-fold $K3 \times \mathbb{C}$.
- A complete proof of Conjectures 5.4 and 5.5 (including imprimitive classes) appears in [61]. The geometry of $K3$-fibered Calabi–Yau 3-folds plays an essential role.

Conjecture 5.3 remains open in the imprimitive cases.

Steps beyond Conjectures 5.4 and 5.5 have been proposed in [29]: a full motivic lift of the Katz–Klemm–Vafa formula for stable pairs that is related to the generating series of Hodge numbers of Hilbert schemes of points of $K3$ surfaces. For motivic invariants of the stable pairs theory on $K3$ surfaces, many phenomena remain mysterious.

Notes

1. All varieties here are defined over $\mathbb{C}$.
2. Usually $\overline{M}_g(X, \beta)$ denotes the moduli space of stable maps with connected domains. The bullet in our notation indicates the possibility of disconnected domains. We always require the maps to be non-constant on each connected component of the domain.

3. We assume, for the given interpretation of terms, that the domain C has no infinitesimal automorphisms.

4. Sometimes $Z_{GW,\beta}(u)$ as defined here is called the *reduced* partition function since the constant map contributions are absent. The constant contributions, calculated in [14], will not arise in our discussion.

5. $[F]$ is the sum of the classes of the irreducible 1-dimensional curves on which F is supported weighted by the generic length of F on the curve.

6. Every fixed-determinant deformation of the complex (to any order) is quasi-isomorphic to a complex arising from a flat deformation of a stable pair [58]. However, the obstruction theory obtained from derived category deformations differs from the classical deformation theory of pairs.

7. A local Calabi–Yau toric surface is the total space of the canonical bundle of any non-singular, projective, toric Fano surface.

8. In [46], the most general local Calabi–Yau toric geometry involving the 3-leg vertex is analyzed for the Gromov–Witten/Donaldson–Thomas correspondence. The same path of argument applies to stable pairs theory also [50].

9. The obstruction theory is equipped with a pairing identifying the tangent space with the dual of the obstruction space [3].

10. The integral is defined by

$$\int_{P_n(X,\beta)} \chi^B = \sum_{n \in \mathbb{Z}} n \chi_{\text{top}}((\chi^B)^{-1}(n)),$$

where χ_{top} on the right is the usual Euler characteristic.

11. A class β is irreducible if all 1-dimensional subschemes representing β are reduced and irreducible.

12. By Poincaré duality, there is a canonical isomorphism $H_2(S,\mathbb{Z}) \cong H^2(S,\mathbb{Z})$, so we may view curve classes as taking values in either theory.

13. The group of isometries of the K3 lattice $U^3 \oplus E_8(-1)^2$ acts transitively on elements with fixed norm and divisibility. The dependence of the reduced Gromov–Witten on only the norm and divisibility then follows from the global Torelli theorem. See [10] for a slightly different point of view on the same result.

14. See, e.g., Lemma 2 of [49].

15. The Eisenstein series E_{2k} is the modular form defined by the equation

$$-\frac{B_{2k}}{4k} E_{2k}(q) = -\frac{B_{2k}}{4k} + \sum_{n \geq 1} \sigma_{2k-1}(n) q^n,$$

where B_{2n} is the $(2n)$th Bernoulli number and $\sigma_n(k)$ is the sum of the kth powers of the divisors of n,

$$\sigma_k(n) = \sum_{i|n} i^k.$$

16. Our indexing conventions differ slightly from those adopted in [10].

17. BPS state counts can be extracted from Gromov–Witten theory via [21, 22]. The counts $r_{g,\beta}$ are conjecturally *integers*.

18. The conflicting uses of q seem impossible to avoid. The possibilities for confusion are great.
19. Independence of m holds when $2m^2$ divides $2h - 2$. Otherwise, no such class β exists and $r_{0,m,h}$ is defined to vanish.
20. A class in $H^2(S, \mathbb{Z})$ of divisibility 1 is *primitive*.
21. An embedding of lattices is primitive if the quotient is torsion-free.
22. While the results of the papers [7, 36] have considerable overlap, we will follow the point of view of Borcherds.
23. If f is a series in q, then $f[k]$ denotes the coefficient of q^k.
24. The result of [49] is stated in the rank $r = 1$ case, but the argument is identical for arbitrary r.
25. The model has been studied in physics since the 1980s. The letter S stands for the dilaton, while T and U label the torus moduli in the heterotic string. The STU model was an important example for the duality between type IIA and heterotic strings formulated in [27] and has been intensively studied [24, 25, 32, 33, 43].

References

[1] M. Aganagic, A. Klemm, M. Marino and C. Vafa, *The topological vertex*, Commun. Math. Phys. **254** (2005) 425–478.

[2] K. Behrend, *Gromov–Witten invariants in algebraic geometry*, Invent. Math. **127** (1997) 601–617.

[3] K. Behrend, *Donaldson–Thomas invariants via microlocal geometry*, Ann. Math. **170** (2009) 1307–1338.

[4] K. Behrend and B. Fantechi, *The intrinsic normal cone*, Invent. Math. **128** (1997) 45–88.

[5] K. Behrend and Yu. Manin, *Stacks of stable maps and Gromov–Witten invariants*, Duke Math. J. **85** (1996) 1–60.

[6] A. Beauville, *Counting rational curves on K3 surfaces*, Duke Math. J. **97** (1999) 99–108.

[7] R. Borcherds, *The Gross–Kohnen–Zagier theorem in higher dimensions*, Duke Math. J. **97** (1999) 219–233.

[8] T. Bridgeland, *Hall algebras and curve-counting invariants*, J. Am. Math. Soc. **24** (2011) 969–998.

[9] J. Bryan and C. Leung, *Generating functions for the numbers of curves on abelian surfaces*, Duke Math. J. **99** (1999) 311–328.

[10] J. Bryan and C. Leung, *The enumerative geometry of K3 surfaces and modular forms*, J. Am. Math. Soc. **13** (2000) 371–410.

[11] X. Chen, *Rational curves on K3 surfaces*, J. Alg. Geom. **8** (1999) 245–278.

[12] I. Dolgachev and S. Kondo, *Moduli of K3 surfaces and complex ball quotients*, arXiv:math/0511051 [math.AG].

[13] S. Donaldson and R. Thomas, *Gauge theory in higher dimensions*, in S. A. Huggett, L. J. Mason, K. P. Tod, S. Tsou and N. M. J. Woodhouse, eds., *The Geometric Universe: Science, Geometry, and the Work of Roger Penrose*(Oxford University Press, 1998), pp. 31–48.

[14] C. Faber and R. Pandharipande, *Hodge integrals and Gromov–Witten theory*, Invent. Math. **139** (2000) 173–199.

[15] C. Faber and R. Pandharipande, *Relative maps and tautological classes*, J. Eur. Math. Soc. (JEMS) **7** (2005) 13–49.

[16] B. Fantechi, L. Göttsche and D. van Straten, *Euler number of the compactified Jacobian and multiplicity of rational curves*, J. Alg. Geom. **8** (1999) 115–133.

[17] W. Fulton and R. Pandharipande, *Notes on stable maps and quantum cohomology*, Proc. Symp. Pure Math. **62**(2) (1997) 45–96.

[18] A. Givental, *Equivariant Gromov–Witten invariants*, Int. Math. Res. Not. **13** (1996) 613–663.

[19] A. Givental, *A mirror theorem for toric complete intersections* in M. Kashiwara, A. Matsuo, K. Saito and I. Satake, eds., *Topological Field Theory, Primitive Forms, and Related Topics* (Birkhäuser, 1998), pp. 141–175.

[20] A. Gathmann, *The number of plane conics 5-fold tangent to a smooth curve*, Compos. Math. **141** (2005) 487–501.

[21] R. Gopakumar and C. Vafa, *M-theory and topological strings—I*, arXiv:hep-th/9809187.

[22] R. Gopakumar and C. Vafa, *M-theory and topological strings—II*, arXiv:hep-th/9812127.

[23] T. Graber and R. Pandharipande, *Localization of virtual classes*, Invent. Math. **135** (1999) 487–518.

[24] J. Harvey and G. Moore, *Algebras, BPS states, and strings*, Nucl. Phys. B **463** (1996) 315–368.

[25] J. Harvey and G. Moore, *Exact gravitational threshold correction in the FHSV model*, Phys. Rev. D **57** (1998) 2329–2336.

[26] D. Huybrechts and R. P. Thomas. *Deformation–obstruction theory for complexes via Atiyah and Kodaira–Spencer classes*, Math. Ann. **346** (2010) 545–569; arXiv:0805.3527 [math.AG].

[27] S. Kachru and C. Vafa, *Exact results for N = 2 compactifications of heterotic strings*, Nucl. Phys. B **450**, 69–89 (1995).

[28] S. Katz, A. Klemm and C. Vafa, *M-theory, topological strings, and spinning black holes*, Adv. Theor. Math. Phys. **3** (1999) 1445–1537.

[29] S. Katz, A. Klemm and R. Pandharipande, *On the motivic stable pairs invariants of K3 surfaces*, arXiv:1407.3181 [math.AG].

[30] T. Kawai and K Yoshioka, *String partition functions and infinite products*, Adv. Theor. Math. Phys. **4** (2000) 397–485.

[31] B. Kim, A. Kresch and Y.-G. Oh, *A compactification of the space of maps from curves*, Trans. Am. Math. Soc. **366** (2014) 51–74.

[32] A. Klemm, M. Kreuzer, E. Riegler and E. Scheidegger, *Topological string amplitudes, complete intersections Calabi–Yau spaces, and threshold corrections*, JHEP **05** (2005) 023.

[33] A. Klemm, W. Lerche and P. Mayr, *K3-fibrations and heterotic type II string duality*, Phys. Lett. B **357** (1995) 313–322.

[34] A. Klemm, D. Maulik, R. Pandharipande and E. Scheidegger, *Noether–Lefschetz theory and the Yau–Zaslow conjecture*, J. Am. Math. Soc. **23** (2010) 1013–1040.

[35] M. Kontsevich, *Enumeration of rational curves via torus actions*, in R. Dijkgraaf, C. Faber and G. van der Geer, eds., *The Moduli Space of Curves* (Birkhäuser, 1995), pp. 335–368.

[36] S. Kudla and J. Millson, *Intersection numbers of cycles on locally symmetric spaces and Fourier coefficients on holomorphic modular forms in several complex variables*, Pub. IHES **71** (1990) 121–172.

[37] J. Le Potier. *Systèmes cohérents et structures de niveau*, Astérisque **214** (1993).

[38] J. Lee and C. Leung, *Yau–Zaslow formula for non-primitive classes in K3 surfaces*, Geom. Topol. **9** (2005) 1977–2012.

[39] J. Lee and C. Leung, *Counting elliptic curves in K3 surfaces*, J. Alg. Geom. **15** (2006) 591–601.

[40] J. Li and G. Tian, *Virtual moduli cycles and Gromov–Witten invariants of algebraic varieties*, J. Am. Math. Soc. **11** (1998) 119–174.

[41] B. Lian, K. Liu, and S.-T. Yau, *Mirror principle I*, Asian J. Math. **4** (1997) 729–763.

[42] C.-C. Liu, K. Liu, and J. Zhou, *A formula of 2-partition Hodge integrals*, J. Am. Math. Soc. **20** (2007) 149–184.

[43] M. Mariño and G. Moore, *Counting higher genus curves in a Calabi–Yau manifold*, Nucl. Phys. B **453** (1999) 592–614.

[44] D. Maulik, N. Nekrasov, A. Okounkov and R. Pandharipande, *Gromov–Witten theory and Donaldson–Thomas theory. I*, Compos. Math. **142** (2006) 1263–1285.

[45] D. Maulik, N. Nekrasov, A. Okounkov and R. Pandharipande, *Gromov–Witten theory and Donaldson–Thomas theory. II*, Compos. Math. **142** (2006) 1286–1304.

[46] D. Maulik, A. Oblomkov, A. Okounkov and R. Pandharipande, *Gromov–Witten/Donaldson–Thomas correspondence for toric 3-folds*, Invent. Math. **186** (2011) 435–479.

[47] D. Maulik and R. Pandharipande, *A topological view of Gromov–Witten theory*, Topology **45** (2006) 887–918.

[48] D. Maulik and R. Pandharipande, *New calculations in Gromov–Witten theory*, Pure Appl. Math. Q. **4** (2008) 469–500.

[49] D. Maulik and R. Pandharipande, *Gromov–Witten theory and Noether–Lefschetz theory*, in B. Hassett, J. McKernan, J. Starr and R. Vakil, eds., *A Celebration of Algebraic Geometry* (Clay Mathematics Institute/American Mathematical Society, 2013), pp. 469–507.

[50] D. Maulik, R. Pandharipande and R. Thomas, *Curves on K3 surfaces and modular forms*, J. Topol. **3** (2010) 937–996.

[51] A. Okounkov and R. Pandharipande, *Gromov–Witten theory, Hurwitz numbers, and completed cycles*, Ann. Math **163** (2006) 517–560.

[52] A. Okounkov and R. Pandharipande, *The equivariant Gromov–Witten theory of* $\mathbf{P}^1$, Ann. Math **163** (2006) 561–605.

[53] A. Okounkov and R. Pandharipande, *Virasoro constraints for target curves*, Invent. Math. **163** (2006) 47–108.

[54] R. Pandharipande and A. Pixton, *Descendents on local curves: rationality*, Compos. Math. **149** (2013) 81–124.

[55] R. Pandharipande and A. Pixton, *Descendent theory for stable pairs on toric 3-folds*, J. Math. Soc. Japan **65** (2013) 1337–1372.

[56] R. Pandharipande and A. Pixton, *Gromov–Witten/pairs descendent correspondence for toric 3-folds*, Geom. Topol. **18** (2014) 2747–2821.

[57] R. Pandharipande and A. Pixton, *Gromov–Witten/pairs correspondence for the quintic 3-fold*, J. Am. Math. Soc. (to appear)..

[58] R. Pandharipande and R. P. Thomas, *Curve counting via stable pairs in the derived category*, Invent. Math. **178** (2009) 407–447.

[59] R. Pandharipande and R. Thomas, *The 3-fold vertex via stable pairs*, Geom. Topol. **13** (2009) 1835–1876.

[60] R. Pandharipande and R. P. Thomas, *Stable pairs and BPS invariants*, J. Am. Math. Soc. **23** (2010) 267–297.

[61] R. Pandharipande and R. Thomas, *The Katz–Klemm–Vafa conjecture for K3 surfaces*, Forum Math. Pi **4** (2016) e4.

[62] A. Pixton, *The Gromov–Witten theory of an elliptic curve and quasi-modular forms*, Senior thesis, Princeton, 2008.

[63] R. Thomas, *A holomorphic Casson invariant for Calabi–Yau 3-folds and bundles on K3 fibrations*, J. Differential Geom. **54** (2000) 367–438.

[64] Y. Toda, *Generating functions of stable pair invariants via wall-crossing in the derived category*, Adv. Stud. Pure. Math. **59** (2010) 389–434.

[65] B. Wu, *The number of rational curves on K3 surfaces*, Asian J. Math. **11** (2007) 635–650.

[66] S.-T. Yau and E. Zaslow, *BPS states, string duality, and nodal curves on K3*, Nucl. Phys. B **457** (1995) 484–512.

Index